THE TECHNIQUE OF THOUGHT

THE TECHNIQUE OF THOUGHT

Nancy, Laruelle, Malabou, and Stiegler after Naturalism

IAN JAMES

University of Minnesota Press
Minneapolis
London

The University of Minnesota Press gratefully acknowledges the financial assistance provided for the publication of this book by the Modern and Medieval Languages Faculty at Cambridge University.

Portions of chapter 4 were previously published as "(Neuro)-plasticity, Epigenesis, and the Void," *Parrhesia* 25 (2016): 1–19.

Copyright 2019 by the Regents of the University of Minnesota

All rights reserved. No part of this publication may be reproduced, stored in a retrieval system, or transmitted, in any form or by any means, electronic, mechanical, photocopying, recording, or otherwise, without the prior written permission of the publisher.

Published by the University of Minnesota Press
111 Third Avenue South, Suite 290
Minneapolis, MN 55401-2520
http://www.upress.umn.edu

Printed in the United States of America on acid-free paper

The University of Minnesota is an equal-opportunity educator and employer.

Library of Congress Cataloging-in-Publication Data

Names: James, Ian, author.

Title: The technique of thought : Nancy, Laruelle, Malabou, and Stiegler after naturalism / Ian James.

Description: Minneapolis : University of Minnesota Press, 2019 | Includes bibliographical references and index. |

Identifiers: LCCN 2018022886 (print) | ISBN 978-1-5179-0429-6 (hc) | ISBN 978-1-5179-0430-2 (pb)

Subjects: LCSH: Philosophy, French—20th century. | Philosophy, French—21st century. | Nancy, Jean-Luc. | Laruelle, François. | Malabou, Catherine. | Stiegler, Bernard.

Classification: LCC B2421 .J3645 2019 (print) | DDC 194—dc23

LC record available at https://lccn.loc.gov/2018022886

UMP BmB 2019

FOR ANDREW JAMES AND SALLY HOWE

CONTENTS

PREFACE

ALTHOUGH IT HAS ITS ORIGINS IN RESEARCH carried out in the middle of the previous decade, this book was conceived and written between 2012 and 2017. This period witnessed the emergence of what has been called the "post-truth era." As shown by the rise of Donald Trump in the United States and the (narrow) vote in favor of the United Kingdom leaving the European Union, the emergence of this crisis of public and political debate has had consequences that have yet to be fully resolved at the time of this writing. The root causes of this phenomenon have sometimes been located in the economic and social exclusion that has been the legacy of the 2008 global financial crash. The nature of modern digital media and the rise of social networking have also been blamed for the loosening of the bonds between public and political discourse on the one hand and verifiable truths or bodies of scientific and evidence-based knowledge on the other.

Some have laid the blame, at least in part, on the philosophical and critical discourses that arose largely, but by no means exclusively, in France in the last decades of the twentieth century and that were grouped together under the banner of postmodernism. In overturning philosophical foundations, the argument goes, and in questioning the self-identity and unity of philosophical concepts or knowledge structures, such discourses opened the door for widespread epistemic relativism, a disregard for facts and truth, and a general undermining of objectivity. The reality of the situation such as it has developed is unquestionably more complex than such arguments admit, in part because debate in this area has often been entirely divorced from any knowledge of the original philosophical texts and contexts in question. The transmission of French and European philosophical discourses into those of the Anglophone academy across various disciplines will certainly be an important subject for future intellectual historians, just as the wider phenomenon of the post-truth era will be for historians more generally. For now, though, it seems that traditional modes of knowledge, and our theoretical or philosophical categories, no longer allow us to fully understand this situation.

What is clear, however, is that a rift appears to be opening up between what the great American naturalist Wilfrid Sellars called the "scientific" picture of the world and the "manifest" pictures of the world belonging to the multiple social, economic, and political communities of North America, Europe, and beyond. Sellars's dream of a unification of scientific understanding with the everyday worldviews of diverse human communities appears to be far from its full realization. This comes at a time when global environmental and security challenges are more pressing than ever and when a sense of crisis continues to make itself felt on all fronts. With climate change denial firmly entrenched in our politics and the return of far-right discourses into the mainstream, this sense of global crisis is intimately bound up with the crisis of post-truth culture and the widening rift between the scientific and manifest world pictures.

The aims and scope of philosophical thought are necessarily modest in relation to such a complex situation. Nevertheless, Sellars's ambition for philosophy to have some role to play in bringing together the scientific picture of the world with ordinary conceptions and everyday worldviews is perhaps more relevant today than ever. In bringing postphenomenological and postdeconstructive philosophical discourses into dialogue with scientific perspectives, thought, and knowledge, this book reflects a similar ambition. Despite many claims in Anglophone debates concerning French theory in general and deconstruction in particular, there is nothing in this book that seeks to undermine or relativize the validity of scientific knowledge. Indeed, it proposes a strong, albeit novel and variably configured, scientific realism that is compatible with a novel and variably configured naturalism. In this context, philosophy of science debates that oppose, say, realism to constructivism are superseded. Most importantly, phenomenological and postphenomenological perspectives that deal with individual or collective consciousness and first-person qualitative experience are brought and thought together with scientific perspectives that are concerned with objective, physical, biological, and third-person quantitative knowledge. The post-Continental naturalism proposed here allows for the interactions between the scientific and manifest pictures of the world to be rethought in entirely new terms.

The arguments that I make in this book are based in close reading, interpretation, and further development of philosophical texts drawn from a variety of traditions. I aim to be both as clear as possible in the presentation of these texts and to make no assumptions about familiarity with the material covered. I hope that those familiar with Anglophone naturalism, science, or philosophy of science writing but unfamiliar with French thought might read this book. By the same token, it is assumed that those deeply schooled in French and Continental philosophy but less well schooled in philosophy of science or naturalist debates will also be readers. I have tried to make the readings and arguments accessible; I therefore beg the patience of those for whom some of the material is all too familiar.

All works referred to in this volume are cited in English or their English translation. Where published English translations exist, they have been cited by first giving the reference to the original French edition and then to the translation. Untranslated texts are cited in my own translations (the titles given in French with a translation in square brackets at first mention) with page references to the original French editions.

Thanks are due to all those who have helped and supported during the writing of this book. In particular I'd like to thank Philip Armstrong, Cornelius Borck, Martin Crowley, Rocco Gangle, Irving Goh, Laurens ten Kate, John Ó Maoilearca, Harold Pashler, Mauro Senatore, Anthony Paul Smith, and Emma Wilson. Special thanks are due to Gerald Moore, with whom I have shared the exploration of philosophy, science, and technics over the past decade. I am also grateful to the University of Cambridge and to Downing College for the period of extended research leave in 2015, which allowed me to write the core of this book. The warmth, kindness, and generosity of Jean-Luc Nancy, François Laruelle, Catherine Malabou, and Bernard Stiegler have proved to be invaluable throughout the development of this project, and I owe them all a debt of gratitude. I also thank the late Christopher Johnson, whose pioneering writing on French thought and philosophy has left an indelible mark on my work, as it will have done on many others. Special thanks go to both Aurélien Barrau and Anne-Françoise Schmid for their careful readings of chapters 2 and 3 respectively. I am particularly grateful to Aurélien Barrau for his invaluable indications and suggestions. I would also like to offer warm thanks to Doug Armato for his support for this project, and to Gabriel Levin and Mike Stoffel for their patient editorial work. My gratitude to Ruth Deyermond for her generosity, intellectual stimulus, help, and personal support remains infinite. Last, thanks go to my older brother, Andrew James, who introduced me to the ideas of Einsteinian relativity and quantum physics at an early age, and to my sister, Sally Howe, who invited me at an equally early age to join a school philosophy club. For their role in making this book possible, I dedicate it to them with love.

INTRODUCTION

Post-Continental Naturalism: A Question

TO POSE THE QUESTION of the "technique of thought" is to inquire into the very essence and definition of philosophy itself. It calls into question the way philosophy can, or should, be practiced, and with this the objects of its inquiry, its relation to other forms of knowledge, and even perhaps its ultimate aims or goals. To pose this question is to ask how and why philosophy legislates for its own methods, procedures, and protocols, and how, in so doing, it gives itself an image of what is—or perhaps more importantly what is not—philosophy. This is a question that inescapably evokes the divisions that define so much of twentieth-century philosophy and its separation into the analytic and Continental traditions. What if the division and opposition of these two traditions are largely a matter of technique? If this is so, then we could more easily, along with Richard Rorty, "imagine a future in which the tiresome 'analytic–continental split' is looked back upon as an unfortunate, temporary breakdown of communication" (Rorty in Sellars 1997, 12).

The technique of thought, posed as a question for philosophy, provides the guiding thread of this book as it seeks to elaborate something that bears the name post-Continental naturalism, a term coined by Mullarkey (2006). It will become clear that if there *is* a something that can go by this name, it is not a philosophical movement or school, and even less a single unitary philosophy. Rather, it is a tendency that can be ascribed to the four contemporary philosophers whom I discuss, philosophers working within a distinctly European, and specifically French, tradition. As a tendency, it emerges from the way they each develop a range of different techniques for thought, different images of what philosophy has been and should become, and then engage in different ways with scientific perspectives and with contemporary scientific knowledge or thinking. In so doing, they radically resituate the relation between philosophy and science into an entirely novel series of configurations. On the basis of this multiple reconfiguration of the relation between philosophy and science, I propose a recasting of philosophical naturalism into a new form. In this way, the term will take on a meaning and scope that will be rather different from that which may currently be generally understood by this term within the Anglophone tradition of philosophy, or indeed within its various already existing Continental articulations. A robust and rigorous questioning of the potential possibilities and limits of the term "naturalism" within philosophy more generally is therefore what the book as a whole aims to provoke.

Jean-Luc Nancy, François Laruelle, Catherine Malabou, and Bernard Stiegler can all, in various ways, be described as postdeconstructive thinkers, either because they were close friends, colleagues, or students of Jacques Derrida, or because their work emerges out of a critique of some of the impasses or problems of deconstruction. Given the widespread perception among those less familiar with or opposed to Derrida's work that it is in some way inimical to the scientific worldview and somehow affirms an epistemic relativism of the worst kind, the association of these post-Derridean thinkers with any kind of philosophical naturalism might appear problematic or tenuous at best, and incoherent and untenable at worst. More importantly and substantively, Derridean thought emerges from the critique (or, more properly, the deconstruction) of a post-Kantian, phenomenological, and therefore idealist or transcendentalist tradition, which vigorously opposes itself to naturalism and to naturalistic attitudes (e.g., psychologism, physicalism, and empiricism). Yet as I argued in *The New French Philosophy* (James 2012), in the wake of Derrida, each of these thinkers quite clearly affirms renewed forms of realism and materialism. The argument of this book is that the thought of each is rigorously and coherently compatible with a kind of naturalism also, albeit one that, as post-Continental, is somewhat different from what has been understood as such hitherto.

As David Papineau notes, the term "naturalism," although familiar and widely used, carries with it "little consensus on its meaning" (1993, 1). The more recent entry on naturalism in the *Stanford Encyclopedia of Philosophy* confirms his view, noting that it "has no very precise meaning in contemporary philosophy" and that it is "fruitless to try to adjudicate some official way of understanding the term" (https://plato.stanford.edu/index.html). Despite this, any attempt to give rigor and coherence to a renewed understanding of naturalism will have to respond precisely to the broad tradition such as it evolved in the twentieth century and continues to evolve today. In *Philosophical Naturalism*, Papineau for his part defines the term with reference to three principal traits. First, naturalism carries with it "the affirmation of a continuity between philosophy and empirical science"; second, for many naturalists, "the rejection of dualism is the crucial requirement"; and third, for others, "an externalist approach to epistemology" is essential (1993, 1). To this he adds a further commitment that forms the central preoccupation of his own book: the commitment to the doctrine of physicalism.

The post-Continental naturalism that will be elaborated here renegotiates these three commitments (and by extension the fourth) in new, perhaps surprising ways. As will become clear, the continuity between philosophy and science gives way to a relative autonomy of each with the other, which nevertheless keeps them in a necessary relation of close proximity and reciprocal exchange. Dualism is unequivocally rejected, but the oppositions that naturalist antidualism can be said to presuppose—those of the transcendental and empirical, phenomenal and physical, mind and body—are all resituated in different terms.[1]

As resolutely realist, post-Continental naturalism accepts that knowledge has its origin in the real and by way of a causality or determination of thought by the real. Yet it rejects the simplicity of the internal–external opposition that is implied in naturalism's third essential requirement (according to Papineau): an "externalist approach to epistemology." The relation of post-Continental naturalism to what might be termed traditional naturalism is thus a complex one that needs to be carefully drawn out over the course of this book. The key starting point has to be the first, and perhaps most decisive, trait of naturalism: the continuity between philosophy and science.

The Continuity between Philosophy and Science

The tradition of naturalism at stake here is largely American; it runs from figures such as John Dewey, Frederick Woodbridge, Ernest Nagel, Sidney Hook, and Roy Wood Sellars; through Wilfrid Sellars, W. V. Quine, and David Lewis; to recent and contemporary developments within physicalism, eliminative materialism, and naturalized metaphysics (e.g., Paul

Churchland, Patricia Churchland, David Papineau, James Ladyman, Don Ross, and Harold Kincaid). In relation to the question of the continuity between philosophy and science within this tradition, the positions of Wilfrid Sellars, Quine, and Lewis will be taken as paradigmatic. In "The Influence of Darwin on Philosophy," Dewey declares, "Philosophy foreswears inquiry after absolute origins and absolute finalities in order to explore specific values and the specific conditions that generate them" (1951, 13). In the work of these paradigmatic naturalist thinkers, Dewey's declaration takes on a radical and more systematic form. Yet as will become clear, the continuity between philosophy and science that is affirmed by these thinkers is by no means straightforward.

On the face of it, it would seem that the philosophy–science continuum that naturalism proposes implies a somewhat unequal relation whereby the fundamental ambition of the former is more or less absorbed into the scope and authority of the latter. This may seem to be borne out in one of Sellars's most famous phrases: "In the dimension of describing and explaining the world, science is the measure of all things, of what is that it is, and what is not that is not" (1997, 83). This in turn has implications for the way in which philosophy itself is to be practiced: "The procedures of philosophical analysis as such may make no use of the methods or the results of the sciences. But familiarity with the trend of scientific thought is essential to the *appraisal* of the framework categories of the common-sense picture of the world" (81). To this end, philosophy will need to develop a new language (a new technique, in the terminology of this book) that will allow it to talk about "public objects in Space and Time," a language with its own "autonomous logical structure" that can objectively—that is, in an aperspectival manner that is analogous to the objectivity of the scientific method—explain phenomena (116). Such an approach forms the linchpin of Sellars's fundamental distinction between the scientific and manifest images of the world and his affirmation of "a sense in which the scientific picture of the world *replaces* the common-sense picture; a sense in which the scientific account of 'what there is' *supersedes* the descriptive ontology of everyday life" (82).

In matters of ontology and epistemology, science thus has the first and last word, and naturalist philosophy must submit to the authority of scientific knowledge and be guided in its procedures and technique by the "trend of scientific thought." Yet what might appear to be a straightforward subordination of philosophy to science is, in Sellars at least, far more complex. For if science is the measure of all things "in the dimension of describing and explaining the world," philosophy does other things: it gives normative descriptions through the acts of prescribing and proscribing. More broadly it provides a framework for a synoptic vision that would be "not simply a theoretical unification of scientific understanding with our ordinary conception of the world but also embraces the practical dimensions of human existence" (Delaney et al. 1977, ix). One of the most interesting and original contemporary readers of Sellars, Ray Brassier, notes, "Sellars' rationalistic naturalism grants a decisive role to philosophy. [. . .] Philosophy is not the mere under labourer of empirical science; it retains an autonomous function as legislator of categorical revision" (2014, 112). In the case of Sellars, the continuity between philosophy and science keeps open a space of autonomy in which rational philosophical reflection can proceed independently with respect to empirical scientific knowledge.

Quine's philosophy can perhaps be most readily taken as the paradigm par excellence of the continuity between philosophy and science within naturalist thought. His abandonment of the distinction between analytic and synthetic truths as articulated in 1951 in "Two Dogmas of Empiricism" (collected in Quine 1953) is decisive in this regard. This abandonment, along with that of the other empiricist dogma of reductionism, has, Quine notes, two principal effects.[2] First, and most important, it leads to "a blurring of the supposed boundary between speculative metaphysics and natural science." Second, it instigates "a shift towards pragmatism" (Quine 1953, 20). In this context, "ontological questions [. . .] are on a par with natural science" (45). Epistemology is also grounded in natural science, becoming a branch of psychology such that, as Quine argues in *The Pursuit of Truth*, "The objectivity of our knowledge of the external world remains rooted in our contact with the external world, hence in our neural intake" (1990b, 36). This is the context in which Quine comes to define naturalism in general as "the recognition that it is within science itself, and not in some prior philosophy, that reality is to be identified and described" (1981, 21). Therefore, naturalistic philosophy is "continuous with natural science" and is also "naturally associated with physicalism or materialism" (1990a, 257).

Once again, such formulations might appear unequivocally to affirm a subordination of philosophy to science and to endorse a strongly reductivist scientism. And once again, in matters of ontology and epistemology, science and the scientific method do indeed wield the ultimate authority. However, it is worth noting that unlike some later naturalists, Quine is relatively broad in what he admits under the name of science, including "the farthest flights of physics and cosmology," but also, interestingly, "experimental psychology, history and the social sciences" (1990a, 252). There would be much to debate here regarding the equivalent epistemological grounding that is implied in this formulation between, say, physics and history or the social sciences, and even between history and the social sciences. Yet it is at the very least indicative of the inclusive character of Quine's systematic vision. As he himself puts it: "Scientists and philosophers seek a comprehensive system of the world" (1981, 9). Arguably, this is consistent with his holism, the view that "the unit of empirical significance is the whole of science" and that "the totality of our so-called knowledge or beliefs [. . .] is a man-made fabric which impinges on experience only along the edges" (1953, 42). According to holism, any empirical fact or statement only has meaning in relation to sets of statements and wider knowledge structures and in relation to the horizon of what Quine calls "total science, mathematical and natural and human" (45). Although philosophy in Quine's naturalism is indeed continuous with empirical science and is also bound both ontologically and epistemologically by scientific knowledge, the exercise of philosophical reason clearly has, as it does in Sellars, a function that is not reducible to the content of empirical statements or experimental constructs alone. Philosophy has the role of dealing with theoretical and abstract questions that relate to the coherence and systematicity of the whole or totality of knowledge over and above the observation sentences of experimental research. Bound by the authority of science and the scientific method (which has the task of "specifying how reality really is"; Quine 1966, 219), philosophy nevertheless has a strong margin of independent authority in relation to the coherent ordering of total science—that is, the totality of knowledge in general.

This schematic overview of the continuity between philosophy and science in the naturalist philosophies of Sellars and Quine indicates that in each case, and in different but similar ways, such a continuity by no means entails a relation of slavish subordination of the former to the latter. Rather, philosophy aligns itself with, and orientates itself toward, a horizon of completeness or totality (the synoptic vision in Sellars and total science in Quine) that gives it a specific autonomy and function as philosophy, one irreducible to the sciences and to scientific theories as such. Such an autonomy can be discerned in the thought of the third philosopher taken here to be paradigmatic of naturalism: David Lewis.

Lewis's doctrine of "Humean supervenience" offers perhaps one of the strictest and most reductively materialist theses within the naturalist tradition as outlined here. In many ways the doctrine offers an acid test for any philosophical thinking that would like to call itself naturalist. Humean supervenience closely recalls Sellars's demand for a theoretical language that would rest "on a framework about public objects in Space and Time" (1997, 116) insofar as it claims that all explanations should make reference to objects in time and space in order to qualify as true. As Lewis puts it, it is the thesis that "the whole truth about a world like ours supervenes on the spatio-temporal distribution of local qualities" (1999, 224). This means in consequence that "'how things are' is fully given by the fundamental, perfectly natural, properties and relations that those things instantiate," and therefore also that "we may be certain *a priori* that any contingent truth whatever is made true, somehow, by the pattern of instantiation of fundamental properties and relations by particular things" (225). The doctrine offers a general account of philosophical truth that is ontological in character, namely that "truth is supervenient on being" (25). Being, however, is taken to be the "physical" existence determined by science such that "the whole truth supervenes on physical truth" (293) and that truth in general "must be made true, somehow, by the spatio-temporal arrangement of local qualities" (228).

Quine, Lewis's teacher, had already claimed that naturalism was intrinsically associated with materialism and physicalism, and Humean supervenience offers a radical articulation of this claim, transforming the continuity between philosophical and scientific truth into an apparent total identity of the one with the other in which the two become more or less indistinguishable. Yet once again questions arise. First and foremost is the question of whether Humean supervenience is itself compatible with the fundamentals of physical science as we now know them, and whether it can be taken to be true according to the terms the doctrine sets for itself. Later proponents of a scientific or naturalized metaphysics such as James Ladyman and Don Ross have questioned whether this can be so, given that contemporary quantum mechanics (and specifically the phenomenon of entanglement) overturns the concept of "locality" on which Lewis's doctrine relies (2007, 150). For his part, Lewis insists that Humean supervenience is a "speculative addition" to the general naturalist claim that "truth supervenes on being" (1999, 225), and that what is important is that as a doctrine or argument it has a philosophical value insofar as it is a weapon against nonnaturalist arguments that pose some kind of existence beyond the purely physical (dualism, transcendentalism, phenomenalism, epiphenomenalism, and so on): "I defend the philosophical tenability of Humean supervenience, that defence can doubtless be adapted to whatever better supervenience thesis may emerge from better physics" (226). Although it seeks to align philosophical truth with the truths offered by fundamentals of physical theory, the doctrine also has an autonomous status and utility that relates to its "philosophical tenability."

Once again, an autonomous, independent, or indeed outright speculative dimension of philosophical thinking can be discerned within a resolutely naturalist position that seeks to ground all ontological and epistemological claims or truths within the sphere of scientific knowledge. This is most clearly indicated in the context of Lewis's treatment of *possibilia* in *On the Plurality of Worlds* and in the doctrine of modal realism that he develops in this work. "We have only to believe in the vast realm of *possibilia*," he writes, "and there we find what we need to advance our endeavors. We find the wherewithal to reduce the diversity of notions we accept as primitive, and thereby improve the unity and economy of the theory that is our professional concern—total theory, the whole of what we take to be true" (1986, 4). In order to believe in this vast realm, Lewis proposes that we accept the literal existence of many different worlds existing parallel to our own and entirely inaccessible to it—worlds in which possibilities of our own physical space-time being are actualities in others. "Why believe in the plurality of worlds?" he asks; "because the hypothesis is serviceable, and that is a reason to think that it is true" (3). Put another way: "Modal realism is fruitful; that gives us good reason to believe that it is true" (4); or later, "Modal realism ought to be accepted as true. The theoretical benefits are worth it" (135). The benefits in question—all the benefits of Lewis's powerful modal logic and the plurality of actually existing worlds that accompany it—are benefits for a specific conception of philosophy, theory, and knowledge that is arguably far more philosophical than scientific. The plurality of worlds claim, as a truth statement made by philosophy in this contingent world, cannot be defended as a truth on the basis that it is "made true, somehow, by the spatio-temporal arrangement of local qualities" (1999, 228). Rather, it is simply taken to be true on the basis of an entitlement—one that Lewis assumes as it were philosophically, to "expand our beliefs for the sake of theoretical unity," and on the basis that "if thereby we come to believe the truth, then we obtain knowledge" (1986, 109).

To point this out is not to suggest so much that modal realism and the plurality of worlds claim are in contradiction with the doctrine of Humean supervenience as it is to underline that both doctrines should perhaps be taken as contingent within and immanent to philosophical, rather than scientific, practice. Indeed, Lewis (1999, 226) says as much about Humean supervenience. Supervenience places philosophy in the closest alignment and continuity with science (in particular with physics), and modal realism prescribes a powerful technique for philosophy (modal logic). However, the status of both of these foundational arguments of Lewis's naturalism is ultimately speculative. They are placed in the service of a specific image of philosophy: as ideally articulated in "theoretical unity" and as ordered toward an ideal horizon of "total theory." This image of philosophy is itself philosophical rather than scientific. Although it is true that the ideals of the unity of science, and those of a scientific theory of everything, may well be grounded in a single specific image of the scientific enterprise, it is also transparently the case that (as will become clear in chapters 2 and 3 in particular) such ideals are by no means consensual givens within philosophy of science debates—nor within the current multiplicity of scientific theories and modes of practice.

Standing back from this schematic overview of the continuity between philosophy and science in the work of three paradigmatic and highly influential thinkers of the American naturalist tradition, a number of remarks suggest themselves. It has been argued that this continuity does not in fact imply a relation of subordination of philosophy to science. Rather, philosophy delegates the task of knowing physical reality and of determining what kind of entities can properly be said to exist to science while maintaining for itself the role of regulating, rendering coherent, and ordering the totality of knowledge as such and of orienting the general philosophical and (scientific) theoretical enterprise toward the horizon of a synoptic vision (Sellars), total science (Quine), or total theory (Lewis). There is a relation of mutual benefit and of reciprocal authority bestowal here. Philosophy defers to science's authority concerning knowledge of "how reality really is" (Quine 1966, 219). It also absorbs this authority into its own claims, protocols, and procedures. Strengthened—and ontologically and epistemologically legitimated—by this alignment with and absorption of scientific truth, philosophy at the same time maintains its own authoritative domain: that of regulative and speculative reason, its capacity to stand above and rationally order the currently existing whole of knowledge, and also to maintain the orientation of thought toward a futural horizon of totality or total knowledge. In so doing, it also confers back onto science the all-important imprimatur of philosophical legitimation, making science the cornerstone and first point of reference for knowledge of natural being and existence—that is to say, of all being and existence.

Naturalist Legacies and Counterlegacies

The principal focus here is the relation between philosophy and science within the tradition of naturalism, and the question of whether the renegotiation of this relation by the thinkers discussed in this book has led to the emergence of something like a post-Continental naturalism. Chapter 1 analyzes the different ways in which Nancy, Laruelle, Stiegler, and Malabou propose a series of specific images of philosophy. These images could not be more different from the image that has been discerned in the thought of Sellars, Quine, and Lewis above. First and foremost, they are not derived from any initial appeal to the way empirical science encounters or knows physical reality. Rather, they are derived from encounters with the history of philosophy, from readings of the tradition and historical trajectory of Western thought, and from attempts to articulate a fundamental structure of philosophical thought as such. The image of philosophy they propose arises from an experience of thought itself rather than a reference to the scientific experience or picture of reality. This derivation of an image of philosophy from a reading or interpretation of the history of philosophy itself is a standard procedure of Continental thought, and the texts that are read in this context are drawn from this tradition: Kantian transcendentalism, phenomenology, and existential phenomenology, as well as their subsequent deconstruction. The names of, among others, Kant, Hegel, Nietzsche, Husserl, Heidegger, Derrida, and Deleuze will loom large here.

This is a tradition based on the adventures of a transcendental moment within thought, one that seeks by turns to ground itself, historicize and absolutize itself, render itself existential or immanent, and otherwise deconstruct itself. That this tradition is at best only problematically compatible with naturalism will for many go without saying. Certainly the legacy of naturalist thought that has been bequeathed by Sellars, Quine, and Lewis to recent and contemporary philosophy in the forms of eliminative materialism, naturalized metaphysics, or reductivist physicalism suggests its total incompatibility with the trajectory of posttranscendentalist and postphenomenological philosophy engaged with here. So for instance the eliminative materialism of Paul Churchland and Patricia Churchland (discussed in chapter 4) inherits much from Sellars (Paul Churchland's doctoral supervisor) and seeks, as its name suggests, to eliminate from the sphere of philosophical and ontological truth those dimensions of experience that constitute the central preoccupations of the post-Kantian tradition: the transcendental foundations of knowledge, the first-person perspective of consciousness, and everything associated with the qualitative nature of experience as it is subjectively and intersubjectively lived: qualia, desires, beliefs, and so on. In the same way the naturalized metaphysics, overt scientism, and resolute physicalism of James Ladyman and Don Ross (discussed in chapter 2) and that of David Wallace (discussed in chapter 3) take over, and radicalize further, the continuity between science and philosophy proposed by Quine and Lewis. As will become clear in later discussions, Ladyman and Ross as well as Wallace radically constrain ontology within the bounds of contemporary fundamental physics (and in the case of Wallace within the bounds of a specific interpretation of quantum mechanics). In so doing, they also engage in a radically eliminative gesture with respect to phenomenal experience (eliminating the fundamental existence of objects or things, qualia, or, in the case of Wallace, the experimental perspective that defines the quantum measurement problem). In this context, it would appear to be a straightforward conclusion that the posttranscendentalist and postphenomenological images of philosophy offered by Nancy, Laruelle, Stiegler, and Malabou are in no way compatible with naturalist thinking.

Other contexts, however, suggest that things might not be so straightforward. For instance, a distinct recent and still burgeoning philosophical literature has sought to naturalize phenomenology. One might most obviously cite in this context the work on neurophenomenology by Francesco Varela in *The Embodied Mind: Cognitive Science and Human Experience* (1991) and Humberto Maturana and Varela's *Autopoiesis and Cognition: The Recognition of the Living* (1980). One could mention other thinkers who have tried to combine phenomenological accounts of embodiment with cognitivist approaches, such as Antonio Damasio (1996, 2000, 2010), Mark Wrathall and Sean Kelly (1996), and Hubert Dreyfus (1982).[3] Yet other works such as Shaun Gallagher's and Dan Zahavi's *The Phenomenological Mind* (2012) powerfully argue argued that the phenomenological tradition from Husserl through to Sartre, Merleau-Ponty, and beyond can offer an indispensable resource for thinking through issues that go to the heart of contemporary debates around cognition, consciousness, and the scientific exploration of both.[4] This list is by no means exhaustive, and the varied attempts over recent decades to naturalize phenomenology, and to bring it into dialogue with biology and neuroscience in particular, suggest that the antinaturalism that is clearly present at the beginning of this tradition in its modern twentieth-century trajectory (most obviously in Husserl and Heidegger) is by no means an ongoing given of philosophical debate.

Another important context in this regard is to be found in a number of recent developments within the tradition of Continental philosophy over the past two decades. These include first and most prominently the alignment of Deleuzian thought with scientific perspectives. The development over the last ten years or more of a series of disparate philosophical positions that have gone under the name of "speculative realism" should also be cited. Such Deleuzian and speculative realist developments also overlap to a large extent with the recent emergence of different forms of new materialism in the work of thinkers such as, among others, Jane Bennett, Rosi Braidotti, Manuel DeLanda, Karen Barad, and Quentin Meillassoux.[5]

The relation of the thinking elaborated here to Deleuze and Guattari's conception of philosophy and science is discussed at length in chapter 1. Here it will be clear that post-Continental naturalism is different from anything that can be imagined by way of a Deleuzian naturalism and from the various recent bodies of thought that have been inspired by Deleuze. Through a close reading of *What Is Philosophy?* (Deleuze and Guattari 1991, 1994), I argue that Deleuzian immanence does not allow for the thinking of the limit (or of limits) that is engaged with here in relation to both philosophy and science (specifically in Nancy, Stiegler, and Malabou). There has been a significant focus on the ways in which Deleuzian thought corresponds with already constituted scientific thought and knowledge (e.g., complexity theory) and on how both Deleuzian metaphysics and the scientific understanding of the universe come, in different ways, to similar conclusions about reality (DeLanda 2002, 2). The specific negotiation with limits in relation to both philosophy and scientific thought that is uncovered here leads to a significantly different perspective from that of Deleuze. What is of decisive interest here is the manner in which both philosophy and scientific thought and practice interrogate, or more speculatively engage with, that which is not yet, or indeed not at all, conceptually or theoretically determined or determinable—that is to say, with what lies beyond currently constituted knowledge. Examples of this abound in cosmology (Roberto Mangabeira Unger and Lee Smolin), quantum physics (Bernard d'Espagnat, engaged with extensively in chapter 3), or the biology of the origins of life (Nick Lane). This leads to a different relation of philosophy to science than the one that Deleuze-inspired thought is able to articulate.

The distinctiveness of post-Continental naturalism in relation to the diverse philosophical positions going under the name "speculative realism" is highlighted in an extended discussion of things and objects in chapter 2. Here the work of Graham Harman offers a key point of reference and engagement. Harman's object-oriented ontology, with its reliance on a renewed understanding of substance, is contrasted with Nancy's relational ontology, with its reliance on his specific reworking of the term "sense." On this basis, the distinctiveness of Nancy's materialist realism in relation to speculative realism is highlighted and the beginnings of something like a Nancean speculative naturalism discerned. Because many of the thinkers of the new materialism (e.g., Bennett, Braidotti, De Landa) either pass through or are heavily influenced by Deleuze, or can be situated within the speculative realist debate (e.g., Meillassoux), new materialism itself has not been given any separate or extended treatment.[6]

In many ways this book seeks to do for postdeconstructive thought what naturalized phenomenology, Deleuzian naturalism, and new materialism have, in their diverse ways, set out to do: connect the Continental tradition to scientific perspectives and thinking. Yet these already established tendencies make this connection on the basis of different assumptions about the status of philosophy and the experience of thinking; this, I argue, makes all the difference. After establishing in chapter 1 the distinct images of philosophy offered by Nancy, Laruelle, Stiegler, and Malabou, each subsequent chapter examines the way their thinking can be related to contemporary scientific perspectives and the way each explicitly engages with these. In chapter 2, the Nancean philosophy of sense and ontology of singular plurality is first related to debates within contemporary scientific metaphysics about the status of thinghood. Nancean sense is then related to the questioning within biological thought concerning the status of life and living entities. Third and finally, Nancy's ontology is related to questions within cosmology that concern the structure of the universe and the metaphysics of science itself. In chapter 3, the importance of Laruelle's thought for renewing the status of scientific realism, as well as the nature of the scientific enterprise and its relation to wider knowledge, are discussed. Insofar as Malabou and Stiegler share a profound concern with consciousness, individuation and its relation to embodiment, and neuroscientific and technoscientific contexts, they are treated together in chapter 4.

The questions that lie at the heart of philosophical naturalism, the relation of philosophy to science, the rejection of dualism, and the causation of our knowledge of the world in and by the real, are reposed and reconfigured in fundamental ways that remain firmly compatible with something that can still be called naturalist. Yet the horizons of both philosophy and science undergo a radical transformation in this process of reconfiguration. The horizon of totality, unity, and completeness that underpins the images of philosophy found in Sellars, Quine, and Lewis gives way to a horizon of multiplicity, disunity, and incompleteness. Where analytic philosophy found a path back to metaphysics in Quinian and Lewisian naturalism, the post-Continental naturalism elaborated here remains firmly within the closure or deconstruction of metaphysical thought and foundations. It thereby questions the image of science itself that both American naturalism and some scientific theories continue to propose. It also enters into a significant dialogue in its own right with philosophy of science debates relating to realism, instrumentalism, conventionalism, constructivism, and the unity of science. This transformation of horizons has wide-reaching implications for the way in which different areas of knowledge, philosophy, and science, as well as the sciences, social sciences, and humanities, relate to each other.

Above all, post-Continental naturalism allows for the phenomenal and qualitative dimensions of thought and conscious experience (known to phenomenology), and the physicalistic or quantitative dimensions of material existence (known to science) to be brought and thought together without the one seeking to eliminate or otherwise downgrade the other. These two dimensions are thought together in ways that avoid metaphysical totalization or unification, but that also move decisively beyond the scientism, reductivism, and eliminativism that have become the hallmarks of much contemporary philosophy that takes up the legacy of American naturalism. Everything hangs on the way in which the thinkers treated here reconfigure the fundamental image that philosophy proposes for itself as philosophy, and then go on to enter into a different set of dialogues and relations with science and scientific perspectives. As will become clear, this fundamental reconfiguration of the image of philosophy occurs within the context of an irreducible experience of thought, one that engenders and even necessitates innovative and experimental techniques for thought.

1 THE IMAGE OF PHILOSOPHY

GIVEN THAT THE THINKERS DISCUSSED HERE inherit from a tradition that begins with Kant and progresses through Hegelianism, neo-Kantianism, Husserlian phenomenology, existential phenomenology, and then deconstruction, one of the key tasks of elaborating a post-Continental naturalism is to show that the transcendental moment of thought itself is susceptible to naturalization or to being thought in naturalistic terms without simply being abolished or eliminated. This task has already been ongoing in different ways in the contexts of neurophenomenology, Deleuzian engagements with science, and new materialism. Across the chapters that follow, I demonstrate this naturalization of the transcendental to be configured by each of the four thinkers discussed in a singular and distinct manner. It becomes the transimmanence of sense in Nancy, the radical immanence of the Real in Laruelle, the organological conditioning of thought in Stiegler, and epigenetic plasticity in Malabou. In each thinker, these instances are unequivocally material in nature and allow for no ontological duality between thought and matter, between the phenomenal and the physical, between consciousness and the body. Yet within this ontological continuity, the autonomous perspectives of thought, phenomenal perception, or consciousness are preserved rather than reduced or eliminated.

These various postdeconstructive naturalizations of the transcendental do not, however, provide any kind of substantive ontological or metaphysical ground or foundation for thought or being. Nor do they provide any kind of basis on which to secure a sense of unity or totality within thinking. Indeed, what will become clear in the discussions of this and subsequent chapters is that the place of the transcendental within thought and being is void of substance and becomes empty or vacant in a manner that recalls Quine's invocation in *Theories and Things* of "the abyss of the transcendental" (1981, 22). The images of philosophy that are elaborated here in relation to the four thinkers discussed all in various ways articulate an experience of thought that is also an experience of ontological groundlessness, vacuity, or void. One of the key questions I pose in subsequent chapters is the extent to which scientific thinking can similarly affirm the ontological groundlessness or vacuity of physical existence or of the real such as it is known to science. However, it is with thinking itself that the discussion must begin.

Thinking at the Limit: Nancy's *The Experience of Freedom*

Nancy's *The Experience of Freedom* (1988, 1993b) begins with an epigraph taken from the first section of Kant's *Critique of Pure Reason*: "For the issue depends on freedom; and it is in the power of freedom to pass beyond any and every specified limit" (1998, 397 [A317/B374]). As the title of his work suggests, the freedom in question here is not a theme or question to be taken up as the object of philosophical inquiry; rather, it is understood in relation to an experience, and in particular to a specific experience of limits: the limits of philosophy, of subjectivity, and indeed of thought itself. Originally submitted as a dissertation for his doctorate, Nancy's *The Experience of Freedom* is principally an engagement with Kant's thinking of freedom and the key role it plays in critical philosophy and also with Heidegger's ontological understanding of freedom. Its arguments have their origins in the deconstructive and Heidegger-inflected readings of Kant that Nancy elaborates on in his 1970s-era essays, published in works such as *The Discourse of the Syncope: Logodeadalus* (1976, 2007) and *L'Impératif catégorique* [The categorical imperative] (1983). However, and as will become clear, *The Experience of Freedom* (1988, 1993b) is also the first explicit and sustained expression of an ontology of singular plurality or of relational, worldly existence of the kind that receives further development in the 1990s in works such as *The Sense of the World* (1993c, 1997) and *Being Singular Plural* (1996, 2000). As such, it is arguably a pivotal work in Nancy's career as a philosopher, one that carries over the insights of his earlier more commentary-oriented writings of the 1970s and turns them toward the accomplishments of his mature philosophy. Yet it is also a work in which Nancy, perhaps more than anywhere else in his oeuvre, calls into question the very operations and status of philosophy itself. In this way, *The Experience of Freedom* arguably defines the manner in which

Nancy comes to understand philosophy in general—and in particular the manner in which he comes to practice philosophical writing as an experimental gesture that engages the limits of thought as such.

What is perhaps most striking about *The Experience of Freedom* as a whole is the manner in which it explores the close relationship between reason and human freedom in Kantian critical philosophy in distinctively ontological terms.[1] The central role played by reason in the architectonic of Kant's system relates not simply to epistemological concerns (e.g., the manner in which reason serves to unify knowledge) but more fundamentally concerns existence per se. At the beginning of the second chapter of *The Experience of Freedom*, Nancy makes this explicit. Despite the theoretical determination of freedom in Kant's thought, and despite the expectation that critical philosophy will establish it as a principle, Nancy notes that "when freedom was presented in philosophy as 'the keystone of the whole architecture of the system of pure reason' [. . .] what was in question was in fact, and at first, an ostension of the existence of freedom, or more exactly an ostension of its presence at the heart of existence" (1988, 27; 1993b, 21). Rather than being elaborated simply as a principle that will then be closely related to the operation of reason as a faculty, freedom is taken by Nancy to be a fact of reason and then subsequently as a fact of existence (1988, 27; 1993b, 21). It is in this manner that freedom comes to be understood in relation to experience, and specifically in relation to the experience undergone by reason when it encounters or relates to itself in thought or in the discourse of philosophy. In Kant's thought, and with regard to freedom, Nancy writes, "What is involved is the experience that reason *produces* [. . .] from itself" (1988, 28; 1993b, 21–22). This notion of reason encountering or experiencing itself is not new in Nancy's reading of Kant and is central to his problematizing of the foundational ambitions of critical philosophy that forms the arguments of the earlier works *Discourse of the Syncope* and *L'Impératif catégorique* (James 2006, 26–48; Mulqueen and Matthews 2015, 17–32).

What is most significant in this context is that freedom is understood as the freeing from any determinate cause, principle, or foundation that is experienced when reason encounters its own ungroundedness as reason. Freedom for Nancy thus cannot be a thing or an abstract principle (e.g., of autonomy or of the autonomous will) but is rather a fact of reason; it is an absence or withdrawal of the determining ground that reason itself experiences when it tries to relate to, and thereby to ground, itself as reason. Freedom is therefore that which is not founded in any anterior principle, in any antecedent cause, or in any substantive ground of any kind. To this extent, Nancy's freedom resembles its Kantian conception when it is understood as a spontaneity operating in and of itself in the absence of any prior determination or cause (Kant 1998, 533 [A533/B561]). However, it diverges significantly from Kant insofar as it is no longer based on the autonomy of a willing subject. Indeed, as "the very thing that prevents itself from being founded" (Nancy 1988, 16; 1993b, 12), Nancy's freedom comes to mark the outer limit point of the philosophy of the subject insofar as such philosophy would seek a ground of knowledge, thought, and existence in the subject. It thereby also, Nancy argues, marks a certain limit point of philosophy per se. To this extent, his thinking in *The Experience of Freedom* is not only a continuation of the readings of Kant upon which he elaborated in *The Discourse of the Syncope* and *L'Impératif catégorique* but also, albeit more indirectly, a further development of the critique of the philosophy of the subject elaborated on in relation to Descartes in *Ego Sum* (1979, 2016).[2]

Nancy here takes up and continues the Heideggerian theme of the end of philosophy and the Heideggerian thesis that modern philosophy is, more than anything, the philosophy of the subject. From this perspective, the preoccupation with subjectivity, taken as the ground of knowledge and of being (its Cartesian conception in the cogito being both inaugural and paradigmatic), defines philosophical modernity in a decisive manner and articulates the structure of ontotheology within that modernity. The experience of freedom as an absence of ground within reason is therefore also an encounter with the limits of the ontotheological conception of the thinking subject as ground. Nancy describes this limit as "the internal border of the limit of ontotheology," where ontotheology is understood as "absolute subjectivity as the essence of essence, and of existence." He goes on to specify the nature of this limit: "This limit is reached as soon as the logic and signification of *foundation* in general, that is to say philosophy, is reached. The end of philosophy deprives us of a foundation of freedom as much as it deprives us of freedom as foundation; but this 'deprivation' was already inscribed in the philosophical aporia consubstantial with the thought of a foundation of freedom and/or with the thought of freedom as foundation" (1988, 16; 1993b, 12). The stakes of freedom in Nancy's thinking could not be higher. As reason encounters freedom as the withdrawal of any ground or foundation within thought, and as it does so in its very act of self-legislation or self-grounding as reason, it encounters also and in the same moment the limit of any and all logic of foundation in general. It marks the limit and withdrawal of the figure of the subject understood as the ground or essence of being and with that the limit of any philosophy or ontotheology that would hope to thereby establish or disclose a ground or essence of being.

What the experience of freedom in Nancy comes to articulate or disclose is a certain and specific fatality or trajectory of modern European philosophy that unfolds as a result of its ambition to seek a universal ground or foundation for being in the operations of reason or rational subjectivity. It is the decision of philosophy, and the decision of thought as philosophy, to be a discourse of universal foundation and to seek this foundation in the modern period in the both the finitude and freedom of a rational thinking subject, which leads philosophy to its very limit and so to an experience of freedom as the absence of all and any foundation. Nancy describes this in terms of a fatality of philosophy as it comes to touch at the limit of the ontology of the subject:

> If philosophy has reached the limit of the ontology of subjectivity, this is because it has been led to this limit. It was led to that point by the initial decision of philosophy itself. This decision was the decision of freedom [. . .] and it was in any case, and still is, the decision of a freedom necessarily prior to every philosophy of freedom. This was not and is not [. . .] the decision *of* philosophy, but rather the decision *for* philosophy, the decision that delivers and will deliver philosophy to its destiny. (1988, 47–48; 1993b, 7)

This may seem like something of an oblique or opaque formulation. The decision of philosophy to be a thinking of universal (ontological) foundations enacted via the operations of rational subjectivity is a free decision of thought that leads it to the encounter with its own freedom as thought—that is, with freedom experienced in and as the withdrawal of any and all foundation. Freedom, understood as the absence of any foundation within thought, is both that which, anterior to thought, frees thinking for the philosophical decision according to which it seeks to establish a foundation and that which thought encounters at the aporetic terminus of thought to which the philosophical decision is inevitably led and where the absence of universal foundation is experienced. As Nancy puts it: "There is decision for philosophy and philosophical decision to the extent that thinking does not appear to itself in a subject, but receives (itself) from a freedom that is not present to it" (1988, 48–49; 1993b, 8). Nancy's thinking of freedom therefore gives us an image of philosophy in general and as such. In particular, it gives a diachronic image of philosophy, one of a discourse that is led, by way of a certain necessity or fatality, from its initial decision *as* and *for* philosophy to its formulation as the philosophy of subjectivity and then to its exhaustion in an encounter with an outer limit which is constitutive of it as philosophy.

The absence, or withdrawal from presence, of any foundation from philosophical thinking, understood or experienced as the terminus or limit point of the ontology and philosophy of the subject, also necessarily marks the withdrawal of any foundation or ground within being or existence in general. Again Nancy makes this explicit:

> The end of philosophy would be a *deliverance from foundation* in that it would withdraw existence from the necessity of foundation, but also in that it would be set free from foundation, and given over to unfounded "freedom."
>
> At the limit of philosophy, there where we are, not having made our way, but having happened and still happening, there is only [. . .] the free dissemination of existence. This free dissemination [. . .] is not the diffraction of a principle, not the multiple effect of a cause, but is the an-archy—the origin removed from every logic of origin, from every archeology—of a singular and thus in essence plural arising whose being *as being* is neither ground, nor element, nor reason, but truth, which would amount to saying, under the circumstances, freedom. (1988, 16–17; 1993b, 12–13)

This passage highlights, perhaps more than any other passage one might choose from *The Experience of Freedom* (or indeed anywhere else in Nancy's work), the transition from the deconstructive questioning of philosophical foundations that dominates Nancy's commentaries of the 1970s to his mature ontology of singular plural being that becomes the centerpiece of his thought from the early to mid-1990s onward. Here freedom, as the fact or factuality of reason freed from any foundation or ground, is also a fact of existence itself, and therefore frees it from the necessity and logic of metaphysical foundations that rational philosophy has sought for existence. Thus freed, existence is literally nothing other than existence itself. This apparent tautology bears the force of a strong antimetaphysical thinking of being: being can no longer be thought in terms of a unified totality and presence of beings underwritten by an external cause, principle, essence, or ground. There is only, as Nancy puts it, the "free dissemination" of singular existence in the absence of cause, principle, and essence—a dissemination in which singularity emerges necessarily in an irreducible plurality (because each singular is not grounded in any unifying principle) and plurality is therefore always and only the plurality of singularities (because it cannot be subsumed by or circumscribed within any unifying principle).

The image of philosophy that emerges in *The Experience of Freedom* thus combines a deconstructive reading of Kantian rationality with a Heidegger-inflected account of modern philosophy as the philosophy of subjectivity. This is combined with the Heideggerian motif of the end of philosophy to the extent that the encounter of reason with its own freedom (and therefore with its constitutive absence of foundation) is characterized also as an encounter with the limit of ontotheology and thereby with the outer limit point or exhaustion of philosophy as such. At the same time, this image of philosophy opens onto a necessary affirmation of existence as the free dissemination of an ungrounded and uncircumscribable plurality of singularities. It is exactly this complex combination and synthesis of deconstructive and Heideggerian moments that allow Nancy to describe freedom in simple and straightforward terms as "the fact of existence as the essence of itself" (1988, 15; 1993b, 11) and to make bold and apparently sweeping claims, such as, "Freedom perhaps designates nothing more and nothing less than existence itself" (1988, 18; 1993b, 14).

Despite this emphasis on existence, Nancy's discourse is not a description of the withdrawal of philosophical foundations such that philosophy somehow maintains itself as a straightforward ontology. Rather, it has a performative dimension. Philosophy enacts or experiences that withdrawal of foundation and the encounter with the limit in its very activity as thought. In this sense, Nancy's discourse is both a thematization and an activity or practice of thought at its limit. Nancy makes this explicit: "It will be a question of bringing an *experience* of 'freedom' to light as a theme *and* putting it at stake as a *praxis* of thought. An experience is first of all an encounter with an actual given, or rather, in a less simply positive vocabulary, it is the testing of something real" (1988, 25; 1993b, 20). Nancy's thinking traces the trajectory of ontotheology and the philosophy of the subject to the encounter with their limit and the exhaustion of their foundational possibility. Yet he does so in order to experience thought at that limit and thereby to maintain it in that moment of exhaustion. This is so because freedom, as an experience and as a "testing of something real," is not in fact a possible object of thought or something that can be experienced as the content of a subject. As a fact of reason that is encountered only in the withdrawal of foundations, and at the limit of thought, freedom for Nancy can never be an object of thought. Rather, freeing from foundation or ground makes thought possible in the first instance as that very movement of reason able to encounter itself as such. As such, freedom can never in the end be thought but only experienced in and as that experience of the limit where thought encounters its own impossibility of foundation or ground, its own freeing from foundation and ground.

It may seem easy at this point to think that Nancy's thought is turning around a series of repetitions where the motifs of freedom, reason, absence, and limit are ceaselessly reiterated in a manner that builds a thesis more by dint of this reiteration than by any real coherence or consistency of argument. Yet in the end what Nancy is trying to both thematize and perform is the experience of thought held at its limit of possibility and exposed to an impossibility: that of making freedom present or conceptually determinate as such. Crucially, this experience of the limit is not an experience of finitude that marks the limit of what is knowable subjectively of the world as it might be independent of human consciousness or perception. In Nancy's account, this is not an experience of the limit that marks the Kantian distinction between (knowable) phenomena and (unknowable) noumena. What is at stake in the experience of freedom for Nancy is a fundamental and ongoing experience of thought in relation to its ungraspable condition of possibility (freedom); this therefore applies to all thought in general and to philosophical thought in particular. "Freedom," he writes, "makes itself understood, at the limit of comprehension, as what does not originate in comprehension" (1988, 70–71; 1993b, 49). This means that the fatality of thought that Nancy traces in *The Experience of Freedom* becomes, once it has been traced and experienced as such, the site or space in which thought will now knowingly, self-reflexively, and necessarily unfold as a ceaseless encounter with the limits of thought and as an experience of freedom experienced as that which cannot be brought into the orbit and power of the subject. This thinking of thought at the limit, or experience of thought as an experience of a limit, is now the fate of thought itself in its very unfolding or operation as thought: "Every thinking is therefore a thinking about freedom at the same time that it thinks *by* freedom and thinks *in* freedom. [. . .] what happens here, in the free arising of thought, happens precisely on this limit, as the play or very occasion of this limit. Thinking is always thinking on the limit. The limit of comprehending defines thinking" (1998, 75–76; 1993b, 54). Nancy's experience of freedom therefore offers an image of philosophy as a discourse and as a decision of thought. It also offers an image of thought itself. An image of thought as that which continually discovers itself at its limit or experiences itself as unfolding always in an experience at or of the limit of thought: "Thinking thinks the limit, which means there is no thought unless it is carried to the limit of thought" (1988, 76–77; 1993b, 54).

Rather than marking the distinction between a human subject whose knowledge is circumscribed within the realm of the phenomenal and demarcated from an inaccessible noumenal realm, the experience of the limit in Nancy marks an encounter with the constitutive ungroundedness of thought and an encounter with (or more properly an exposure to) the ungrounded and singular plurality of existence as such. As excessive, this singular plurality of existence is not banished into an inaccessible noumenality but rather placed into a relation, albeit one of contact in irreducible distance, with human thought and existence. The realm of singular things, entities, structures, and worldly or cosmological becoming is a surrounding realm with which the singular and ungrounded existence of thought is always in some kind of relation. The nature of this relation will be explored in more detail and in the context of a renewed understanding of scientific knowledge in the next chapter. For now, it is worth noting proleptically that Nancy's experience of the limit as articulated through his reading of Kant does not tie him into any correlationism according to which human thought is bound to its own categories and can only know the relation of correspondence between thought and being. Insofar as it exposes thought to the singular plurality of existence, Nancy's experience of the limit articulates thought itself as that which in a certain way (again to be explored in the next chapter) straddles, exceeds, or otherwise unbinds the correlation between thought and being that Kantianism is said to articulate.

That this limit-experience implicates freedom, thought, and the necessary affirmation of an ungrounded singular plural existence does not mean that Nancy's discourse reverts to ontotheological or metaphysical prejudice at its last and final moment insofar as it continues to speak of existence in general and as such. There is a sense that Nancy's thought persists as something one might call philosophy or ontology, but only barely or minimally. It persists only as this experimental practice of thinking at the limit. One of the key arguments of this book as it situates Nancy's limit-practice of philosophy in relation to science and to scientific knowledge is that his thinking can also be called speculative in a specific sense. Nancy himself does not use the term—and indeed refuses it (1996, 67; 2000, 46). But as will be argued more thoroughly in chapter 2, his thinking of existence as singular plurality and of transimmanent sense can be termed speculative insofar as it persists as a kind of philosophically ungrounded and experimental conjecture or surmise in relation to existence or being. It is a specific kind of conjecture or surmise but not a foundational ontology describing being in the logos of its discourse; nor is it a determination of existence that is in any way empirical or similar to the knowledge of existence that science produces. Insofar as experience plays a decisive role in Nancy's practice of philosophy at the limit, this is in no way the experience of entities or things known to and by the kind of empiricism on which science necessarily draws. It is not experience understood more conventionally as the content of a subject that is at stake here (e.g., sense data, worldly encounters). What is at stake is an experience that exceeds the power of the (rational subject) and attests to its limits and to an excess over those limits (the excess of freedom and of freed existence). Ultimately it is the experience of a failure of thought to locate or ground itself in a transcendental moment.

To reiterate, Nancy's thinking of singular plural existence is not ontology as such; it is not a founded or authoritative discourse on, or logos of, being. It is rather a rather special kind speculative (non)ontology of excess. It is a necessary conjecture or surmise of thought as it experiences the excess of the freedom that frees it *as* thought or as it experiences existence as that which is freed *for* thought. "Ontology," Nancy writes, "has only two formal possibilities. [. . .] Either Being *is* singular (there is only Being, it is unique and absorbs all the common substance of the beingness of beings [. . .] or, *there is no* being *apart from* singularity: each time just this once, and there would be nothing general or common except the 'each time just this once'" (1988, 91–92; 1993b, 66–67). Nancy's thinking of existence as its own essence is ontology only barely and minimally insofar as, in the absence of a singular figure of Being that would ground and unify all beings, the singularity of each instance of existence must be posited as such and posited alongside all other singularities in their nontotalizable plurality. At the same time this singular plurality of existence is always and necessarily experienced as being in excess of the thought that thinks it as singular plurality. Thus, whenever the term "ontology" is used in relation to Nancy's thought, it needs to be qualified in this manner. It needs to be understood as bare or minimal, as in excess of its own discourse or as presented in the withdrawal of what it presents (hence the specific force of "non" in "nonontology" here).

By way of conclusion, and as a means of looking forward to the more extensive treatment of Nancy's transimmanence of sense and its relation to science in the next chapter, two interrelated points must be raised. The first concerns Nancy's minimalist and speculative (non)ontology of singular plural existence. The second relates to the specific image of thought that emerges from his practice of philosophy at the limit. Both these points will be explored at greater length in chapter 2 under the rubric of Nancy's naturalism and the alignment of his discourse with scientific thought and knowledge.

However, first it needs to be noted that Nancy's experience of thought, insofar as it is one in which freedom, existence, and thought itself are intimately bound together, gives rise to an affirmation of an existence not simply as both singular and plural but also, in this singular plurality, as necessarily relational. Insofar as freed existence surges as existence each time singularly in the absence of any ground or foundation, it does so only in relation to other singular instances. "The singular being," Nancy writes, "is in relation, or according to relation, to the same extent that its singularity can consist [. . .] in exempting itself or cutting itself off from every relation" (1988, 91; 1993b, 66). There is a strong sense in which singularity is constitutively relational. A singular instance can only be singular as such when existing in relation to other singularities, which each time surge as existence in the absence of any underlying shared essence, cause, or principle. To this extent, relationality for Nancy is always primordial or anterior to any possibility of severing relation or of existing by way of self-sufficiency or autonomy: "Singularity [. . .] is immediately in relation" (1988, 92; 1993b, 67). At the same time, no singular instance can provide a transcendental condition of possibility or existence for another, just as no singular instance can be entirely immanent to another. They exist, as Nancy would say, transimmanently, each in relation to the other and by way of a relation that knows no ontological substance or ground but only a void or vacuity of substance. The transcendental in Nancy's thought in this sense becomes subject to a logic of a transimmanence that is in excess of any possibility of ontological foundation.

Toward the end of *The Experience of Freedom*, Nancy notes: "There is therefore no thesis here on being except insofar as there is no longer any possible thesis on being. Its freedom is in it and more ancient than it" (1988, 210; 1993b, 167). Philosophy can no longer be a discourse on or of being and its foundation. It can only be the experience of the freedom of being and the aporetic encounter with the limit of thought that this necessitates. "Philosophers," Nancy notes, "have made theses on being; now the question has to do with the fact of its freedom" (1988, 210; 1993b, 167). This is the lesson, or rather the experience of thought, presented in

The Experience of Freedom. It is one that arguably inflects or informs the entirety of Nancy's subsequent thinking. It does so in such a way that we, as readers who may wish to share, reflect on, criticize, or in some novel manner think with Nancy's philosophy, must always take into account the notion that his philosophy is never entirely or exactly philosophy but is rather philosophy brought to its limit and exposed to its ungrounded excess.

Nonphilosophy: Laruelle and Radical Immanence

From the early 1980s onward, Laruelle has, with considerable consistency and rigor, devoted himself to the elaboration of nonphilosophy and to the working out of a theory or a science that, he argues, would be adequate to the radical immanence of the real. In certain respects this brings him close to Nancy's practice of philosophy at the limit, but it also places him in stark opposition to it. Laruelle never mentions Nancy directly, but it is clear that the practice of philosophy at the limit of philosophy simply does not, from the nonphilosophical perspective, go far enough.[3] What has been characterized in relation to Nancy as a thought that is not exactly or entirely philosophy, or is only barely and minimally philosophy, would, according to Laruelle, remain fully within the orbit of the philosophical, or indeed would be the apotheosis of philosophy, the epitome of its essence. The image of philosophy that Laruelle gives is as consistent and rigorous as his characterization of nonphilosophy. As will become clear, his conception of the latter derives precisely and exactly from his structural description of the former. In each of his works from the beginning of the 1980s on, he repeatedly elaborates his structural vision of philosophy in order to repeatedly invent and reinvent nonphilosophical thought in a variety of guises.

Laruelle has the reputation of being the most difficult and impenetrable of contemporary French thinkers. Yet despite this, introductions to and critical presentations of his work tend to be remarkably clear and accessible; they have also grown considerably in number and variety (Mullarkey 2006, 2015; James 2012, 158–80; Gangle 2013; Galloway 2014; Smith 2016a, 2016b; Gangle and Greve 2017). His formal or structural definition of philosophy is, in its consistency and rigor, not difficult to access or understand. More challenging is his definition and theorization of the "non" of nonphilosophy. Yet even that can be given a fairly clear and transparent presentation because it involves a number of axiomatic constants that respond, or are articulated in relation, to the formal structure of philosophy as it is consistently identified across Laruelle's work. What is perhaps most difficult about Laruelle's thought is the challenge posed by any attempt to take the "non" of nonphilosophy seriously, to truly adopt the perspective of the "non," and to effectively avoid treating this thought as just another version of philosophy. This requires the reader to take seriously the change in perspective that his formal and structural image of philosophy proposes and demands. It also requires the reader to take seriously the key distinctions that Laruelle makes between philosophy on the one hand and (nonphilosophical) science or theory on the other. If this change of perspective within thought can be genuinely experienced as such, then even the most obscure or inaccessible aspects of Laruelle's thinking that relate to the "adequation" of nonphilosophical discourse to the immanent real can be revealed or made manifest in a surprising simplicity. The key thing here is never to confuse anything Laruelle might say axiomatically or hypothetically about the immanent Real with ontological or epistemological claims about Being, existence, or the world as it appears to us.

There are many ways to approach the presentation of Laruelle's image of philosophy. These include a simple, formal sketch abstracted from any given Laruellian text and presentations based on one or more specific texts. There have also been attempts to present nonphilosophy by way of discourses or forms that are themselves not philosophical but come from other areas. These include ecology and theology (Smith 2013), film (Ó Maoilearca 2015), or digital media and technology (Galloway 2014). The presentation offered here engages with a number of Laruelle's works spanning the 1980s and 1990s, and it focuses in particular on *Principles of Non-philosophy* (Laruelle 1996, 2013c). However, because my aim here is to bring Laruelle's thought into a relation of comparison and contrast with that of Nancy, I begin by turning toward the critique of Laruelle's immediate predecessors and contemporaries that is provided in *Philosophies of Difference* (Laruelle 1986, 2010b).

Broadly speaking, Laruelle views the philosophy of difference as emblematic of the structure of philosophy as a whole. This includes Derridean deconstruction as well as all those philosophies, including Nancy's, that would engage with a play at the limits of thought or that would seek to somehow affirm an exteriority or excess over thought, the transcendence of an Other that would be irreducible to the Same. In *Philosophies of Difference*, Laruelle's ambition is to elaborate a nonphilosophical mode of thought that is based on a critique of difference—or as he himself puts it: "We seek a non-philosophical critique of philosophy in general and of Difference in particular" (1986, 169; 2010b, 152). This ambition hinges on Laruelle's argument that there is something about the structure of difference that exactly replicates the structure of philosophy as a whole. This argument takes two hypotheses as its starting point—hypotheses that I will explore and elaborate further as this discussion progresses. The first is that "transcendence" is "the inaugural and supreme operation of philosophy." The second is that "the real does not tolerate any operation and is not an operation itself" (1986, 173; 2010b, 155). These hypotheses can be reformulated in even more simple or straightforward terms. The first hypothesis says simply that philosophy as such is always, and necessarily, a capture, a mirroring or representation of the real. It articulates a reflection of the real in and through the transcendence of conceptual determination. The second hypothesis says, again quite straightforwardly, that the real is prior, or antecedent, to any operations of transcendence that conceptual thought or determination might bring to bear on it and that it is entirely indifferent or resistant to such operations. In one way or another, these are two simple hypothetical constants that run throughout Laruelle's nonphilosophical thinking; I elaborate further on their force and specificity in what follows.

The point for Laruelle about difference and the philosophy of difference is that they still offer an image or conceptual determination of the real as or by way of the figure of difference. However much philosophy speaks of the ungroundedness of Being in the plural and the differential flux of becoming (Nietzsche), the withdrawal of Being within ontological difference (Heidegger), or of the deconstruction and dissemination of presence and logos within the play of *différance* (Derrida), it nevertheless remains fully and quintessentially philosophy insofar as it continues to relate the multiplicity or alterity of that which speaks to the unity of a conceptual, and therefore representational, horizon. Laruelle argues that the concept of difference formally and structurally always articulates a division, or a splitting into two or more elements (e.g., logical difference, the differences of multiplicities or of the alterity of temporal-spatial dissemination, the play of differences on plane of immanence). However, as representation, the logical difference, multiplicity, dissemination, or alterity of which difference speaks is always relayed back to, and synthesized with, a unified horizon of conceptual transcendence. As Laruelle puts it, "Difference is Scission-immediately-as-Unity" (1986, 39; 2010b,

24). If this is so, then it means that the ambitions of difference philosophy to affirm a closure of metaphysics and an ontological ungroundedness by way of multiplicity, dissemination, alterity, and so on are always betrayed by the discursive and conceptual operations of difference philosophy itself. Difference philosophy simply never goes far enough in its ambition to deconstruct or overcome the philosophical tradition from which it springs (1986, 175; 2010b, 158). One can see quite easily here how Laruelle, if he were to engage in a critique of Nancy, would without difficulty assimilate his practice of philosophy at the limit to this logic of difference. For Laruelle, Nancy's thinking would seek to affirm an excess beyond the limit, an alterity of relational existence, but that play at the limit of thought would always work to return excess to thought and to its conceptual-representational horizon. This Laruellian reading of Nancy is misplaced, and it belies (or fails to do justice to) the exactitude of the unbinding of thought and its correlation with the real of existence that takes place in Nancy. The crux of the issue here is Laruelle's characterization both of philosophy in general and of difference philosophy in particular as a mode of mixing, or their character as a blending or synthesis of transcendence and immanence (*mixte* in French). It is the image of philosophy in general as a structural *mixte*, critiqued so explicitly in the more local setting of *Philosophies of Difference*, that dominates the entirety of Laruelle's broader nonphilosophical corpus.

Laruelle is consistent in his presentation of the invariant structure of philosophy from the early 1980s onward. Reading his books, from *Le Principe de minorité* [The principle of minority] (1981) through to major late works such as *Philosophie Non-standard* [Nonstandard philosophy] (2010), one has a strange sense of a huge variety of theme and rhetorical gesture combined with an intense sameness or repetition of formal structure. Whether he is elaborating on nonphilosophy proper (Laruelle 1986, 2010b; 1989, 2013b; 1991; 1996, 2013c), a theory of the "ordinary man" (Laruelle 1985, 2017), a theory of Marxism (Laruelle 2000, 2015), a theory of scientific identity (Laruelle 1992, 2016), of ethics (Laruelle 1995), or a heretical nontheology (Laruelle 2002, 2010a; 2014), Laruelle is always presenting a repetition or reinvention of the nonphilosophical in the light of an unchanging characterization of the philosophical. It has been argued that Laruelle's general or all-embracing approach to the definition of philosophy owes too much to Heidegger's characterization of the tradition of Western thought as ontotheology (Brassier 2007, 121). However, Laruelle is far more indebted to structuralist thought, in particular to Louis

Althusser's scientific structuralist approach (James 2012, 163–64). This structuralist dimension to Laruelle's thinking is essential to recognize if one is to do justice to his characterization of philosophy and if his use of the term "science," the central preoccupation of this discussion, is to be fully appreciated.

Laruelle's definitions of the invariant structure of philosophy are given most clearly in works such as *Philosophy and Non-philosophy* (1989, 2013b), *En tant qu'un* [As one] (1991), and *Principles of Non-philosophy* (1996, 2013c). For instance, in *Principles of Non-philosophy,* he speaks of "the most universal trait of philosophy." This invariant feature constitutes philosophy as a discourse that "gives itself an interiority and an exteriority, an immanence and a transcendence *simultaneously,* in a synthetic or hierarchical structure, the one overcoming the other in turn" (1996, 5; 2013c, 4). Therefore, what philosophy does is pose an interiority or immanence to be known on the one hand (the real, being, existence) and the exteriority or transcendence of a figure, concept, or representation of that interiority on the other. When it speaks of any aspect of being or existence, or when it speaks of the real as such, the result is a mixing or synthesis of these two moments of interiority and exteriority, immanence and transcendence. Here the ideality of any concept or representation is always seen as the negation of the immanent real and its sublation into a higher (that is, transcendent) level of abstraction. This is a broadly Hegelian understanding of conceptual determination, and for Laruelle, it carries with it two unfortunate consequences. First, the negativity inherent to the operation of philosophical conceptuality amounts to a kind of violence done unto the real. Second, the resulting synthesis, fusion, or *mixte* of transcendence and immanence, exteriority, and interiority is a transformation or appropriation—that is, a co-constitution of the real in the form of known Being or existence that effectively enacts the violent alienation of the real from itself. The question arises here as to whether this characterization of philosophy, with its distinctly Hegelian flavor and lineage, is not itself necessarily philosophical.

What these characterizations of the structural invariant of philosophy suggest is that philosophy will always repeat its deep structure as philosophy, whatever the form it may take—whether it be a full-blown metaphysics or ontology; any form of idealism, rationalism, positivism, empiricism, or naturalism; or destructive or deconstructive antimetaphysics or forms of skepticism and relativism. This is not so much because of what philosophy says of being, or how it says or speaks of being. Rather, it is so because it is positioned from the outset in what Laruelle calls the philosophical decision. The philosophical decision is not so much a matter of the content of discourse, or a result of the agency or subjectivity of a given philosopher; rather, it is a kind of stance or posture that the philosopher automatically adopts by dint of philosophizing. Again, Laruelle characterizes the philosophical decision according to the structural invariant of philosophy, as here in *Philosophy and Non-philosophy*: "The structural rule of philosophical decision—which is also a transcendental rule, because it is itself affected by affecting what it organises and distributes—is Unity-of-contraries, the circular coextension—with some shifting here and there—of the One and the Dyad" (1989, 12; 2013b, 6–7). The philosophical decision is that stance or posture in which philosophy always already finds itself from the moment it begins to take place as such: the positing of a real to be thought or known, of conceptual thought (of any kind) as the vehicle for such knowledge, and of the mixing or synthesis of both in the resulting philosophical discourse itself. Philosophy and the philosophical decision divides the one (real) into two (the dyad of immanence and transcendence, interiority and exteriority), and unites this two into a greater one (that is, the philosophical determination, representation, or image of the real as Being or as known existence). This, for Laruelle, is philosophy's "transcendental claim to primitively know the real" (1989, 101; 2013b, 99).

The problem here from the Laruellian perspective is not just that philosophy may always enact a violent appropriation of the real or (co-) constitute it in a forced operation of alienation from itself. It is also that philosophy bestows upon itself the authority to do so; it positions itself as the primary authority for the so-called primitive knowledge of the real or of being and thereby legislates for its own ability to legislate over knowledge in general. The structure of splitting or division and subsequent mixing or synthesis has a kind of circularity insofar as it allows philosophy to legislate for its own power to legislate: "Through this structure, philosophy claims to determine itself beyond all its empirical determinations which it only calculates in order to prescribe it in an *auto-position* in which it is titular, an auto-comprehension or auto-legislation, auto-naming etc." (Laruelle 1996, 5; 2013c, 4). Laruelle views the circular nature of philosophy's self-legislation to be a sign of its inherently authoritarian character. The dyadic structure of the philosophical decision allows philosophy to (auto)position itself as the arbiter or legislator of all knowledge of the real, and thereby to exert authority over all other forms of knowledge.

This has two key consequences for Laruelle. First, any one enactment or instantiation of the philosophical decision (e.g., any given philosophy such as empiricism, idealism, positivism) will set itself up as superior to and more "truthful" than any other enactment of the philosophical decision (i.e., all rival philosophies). Philosophy is therefore inherently conflictual and polemical as well as authoritarian. Second, the circularity or arbitrariness of the philosophical decision, no matter how it is enacted, means that all philosophies are equal in relation to the real. They all, in one way or another, enact the structure of splitting, division, and fusion proper to the philosophical decision but equally all are no nearer to doing justice to the radical immanence of the real such as it is hypothesized by Laruelle. All philosophies, of whatever kind, are therefore equally deluded in their pretension to know, seize, or grasp the real. The knowledge, concepts, or representations that philosophy gives us are, in relation to the real at least, entirely illusory.

So Laruelle offers a double characterization of philosophy on the one hand as division, splitting, and unity by way of conceptual transcendence and of the real on the other hand as that which is indivisible and therefore entirely resistant to the operations of conceptual transcendence. This provokes a series of obvious and skeptical questions. By what authority does Laruelle come to such a double characterization? Or, more pointedly, can the authority he necessarily invokes in order to legitimate such a double characterization of philosophy and the real be anything other than a philosophical authority? Does he not inevitably invoke and affirm the very authority he aims to critique and move beyond? These are questions that most readers will ask of Laruelle when they first come to be acquainted with his thinking and many will come to the conclusion that he does indeed remain reliant on claims in relation to both philosophy and the real which cannot but be philosophical in nature. This, precisely, is the difficulty alluded to above of taking seriously the "non" of nonphilosophy and of adopting the perspective Laruelle proposes. And for many this difficulty will be insurmountable.

Along with the question of whether defining philosophy is not always necessarily a philosophical gesture comes the question of whether it is ever really possible to do so at any level of sufficient generality or universality. Anne-François Schmid, a philosopher of science and close collaborator with Laruelle, suggests that it might not be possible to give a general definition of philosophy at all "because any definition would arise within a particular, and therefore partial, philosophy" (2012a, 16). So if defining philosophy is itself always a philosophical gesture, it is always also an impossible one because such a definition will express only the view of one particular philosophy about philosophy and never abstract itself from its own particular position to gain a universal overview of philosophy as such. The only way out of this apparently insuperable impasse would be to concede that it is indeed possible to define philosophy from a perspective that is not philosophical as such but rather theoretical. Schmid puts this helpfully when she argues: "1) one must give oneself the theoretical means for a description of philosophy; 2) one's own opinions and beliefs must be suspended. [. . .] The theoretical means presuppose that philosophy is generalized and that its pretension vis-à-vis the real is removed. That would permit a theoretical (rather than a philosophical) description of philosophy" (2012, 17). The "theoretical means" proposed here would entail a formal definition of the structure of philosophy as that which undeniably has a "transcendental claim to primitively know the real" (Laruelle 1989, 101; 2013b, 99) but that would not itself be philosophical because it suspends that pretension to primitive knowledge. It does not carry the pretension to know the real over either into its definition of philosophy nor into any of its operations as theoretical thinking. This is what Laruelle offers in his characterization of philosophy and of the science of nonphilosophy, and in his understanding of radical immanence as an indivisible real.

It is in this respect that Laruelle's understanding of the real as indivisible and resistant to the operations of conceptual transcendence (i.e., as entirely unknowable) can be viewed—not as a philosophical claim but rather as a working hypothesis or axiom derived from his formal and theoretical description of philosophy. If philosophy always claims primitive knowledge of the real in some form or another, nonphilosophy is simply that which suspends philosophy's claim by treating the real hypothetically as that which cannot be conceptually or representationally known. It is in this context that we need to begin, in preliminary terms, to understand Laruelle's use of the term "science" or "science of the real" to describe nonphilosophy and thus his characterization of the real as "indivisible" radical immanence. In *Philosophies of Difference*, Laruelle defines science in the following terms: "Science is a representation that is non-thetic (of) the real, altogether distinct from what philosophy imagines as Representation" (1986, 177; 2010b, 160). The key to this definition is the use of the term "nonthetic." The *Petit Robert* defines the original French term *thétique* as "that which concerns a thesis" or "that which poses something as existing." So "science" for Laruelle here is that which represents the real without saying anything in general about the real and without binding it into conceptual determination. Science therefore does not offer a thesis on or about the real; it does not name its attributes or make claims about its general being or existence. Philosophy knows the real representationally by dividing it, by forming a *mixte* of immanence and conceptual transcendence, and by offering a synthesis or "thesis" that constitutes its knowledge as such. Science, in Laruelle's understanding of the term, describes the formal structure of the real as that which cannot be divided, mixed, and brought into a synthesis and thesis on or about the real. The characterization of the real as indivisible, or as an unknowable One in Laruelle, thus emerges, in part at least, out of the formal and structural definition of philosophy as a discourse that divides and synthesizes by way of conceptual transcendence. Science here is nonphilosophical, not because it negates philosophy but because it generalizes the structure of philosophy (as division and synthesis) and because it does so nonthetically, suspending the pretensions of that structure to know the real. As Laruelle states in the foreword to *Principles of Non-philosophy*: "the grounding axiom of non-philosophy [is] that the One or the Real is foreclosed to thought and that this is of its own accord rather than owing to a failure of thought" (1996, vi; 2013c, xxii).

In order to take the "non" of nonphilosophy seriously, then, we need simply to take the indivisibility of the immanent real or One seriously as a guiding hypothesis or founding axiom that suspends the pretensions of philosophy, and not take it as a claim made by yet another philosophy. The hypothesis is that the real precedes philosophy or thought, concept, and representation; that the real is entirely indifferent to them; and that the real remains completely untouched by them. It remains undivided by any and all operations of transcendence. This is the simple hypothesis that Laruelle repeats again and again, as here in *Principles of Non-philosophy*: "Thinking the One qua One is thus thinking the radical precession of the One [. . .] over thought, and thinking always the One itself or in-One" (1996, 31; 2013c, 27), or here in the same text: "The One taken up by non-philosophy is neither transcendent nor transcendental; it is only immanent or real, immanent through and through" (1996, 25; 2013c, 22). Laruelle is consistent in his distancing of the One of radical immanence from all the figures of the One that have emerged previously in philosophy (specifically the One of Neoplatonism found in Plotinus, but more generally any horizon of unity such as God, Being, or Reason, because all of these are philosophical conceptual *mixtes* of philosophy). Laruelle's One is one only by dint of its immanence, its absolute and autonomous indifference to thought and representation, its anteriority in relation to all thought and representation.

The ultimate question here is as follows: what happens to philosophy and to thought in general when the philosophical decision is suspended in this manner? Or, put another way, what happens when thought is guided by the hypothesis of the real as an indivisible immanent One? Or again, what can be made to happen and what can be thought or invented within thought when the authority and autopositional self-legislation of philosophy are refused? This in turn would raise the question of the relation of (non)philosophy in this new guise to other areas of knowledge or to knowledge per se. Does nonphilosophy inevitably lead us into a nihilistic skepticism or unbridled and highly corrosive relativism? What becomes of science as understood in the more conventional sense as the natural sciences and the objective knowledge of reality they are widely perceived to offer? Surely the way of nonphilosophy is a path into a kind of theoretical and conceptual delirium. All of these questions and possibilities will be engaged with at length in chapter 3 in the context of a more extended discussion of Laruellian science and its nonepistemological redescription of the natural sciences.

For now, however, it can be said that the image of philosophy that Laruelle presents, if it is indeed to be accepted as a genuinely and effectively nonphilosophical image, is, as it was in Nancy, one of a certain experience of or within thought. The question of whether his nonphilosophy and his theoretical and nonphilosophical description of philosophy is not just more philosophy has been addressed at some length here. I have argued that the "non" is derived from the suspension of the operations of philosophy (division, splitting and synthesis as illusory, violent knowledge of the real) and that this gives rise to the hypothesis of the real as an indivisible One. Yet perhaps more than anything, and in a way that is both similar and different to Nancy's thinking, what all this ultimately amounts to is a certain specific experience of thought and an experimental practice of thinking.

Already in *Philosophies of Difference* Laruelle speaks of the One as a "non-reflexive transcendental experience or absolutely immediate and non-thetic givenness (of) itself" or as a "transcendental experience that is non-thetic (of) itself, absolute and without remainder" (1986, 33, 171; 2010b, 18, 153). Similarly, Laruelle describes the reality of the One as "radical immanence deprived of all transcendence (nothingness, splitting, desire). Immanence lived before all representation as an internal transcendental experience" (1991, 19). Later he describes the first postulate of nonphilosophy as "a type of experience or of the Real which escapes auto-positioning, which is not a circle of the Real and thought" (1996, 6; 2013c, 4). This means that nonphilosophy as such will unfold or be elaborated on the basis of "the real or phenomenal experience [that] can no longer be thought other than through axioms and 'first terms'" (1996, 7; 2013c, 5). These formulations are essential and decisive for the entirety of Laruelle's nonphilosophical project, for what they reveal is that the immanence of the real is not so much framed as a philosophical postulate or claim but rather as something that can be lived and experienced in a certain way, and as such by thought, and only then taken as an initial guiding hypothesis, axiom, or "first term" in a further nonphilosophical experimentation with thought. It may be that Laruelle offers no authority for his project other than this moment of lived immanence or the experience of the real as a nonthetic, indivisible One. There is a sense in which one can simply enter into Laruelle's nonphilosophical gesture of thought by embracing the experience of thought he proposes, along with its resultant hypotheses and axioms, but there is no philosophical authority that would ask us to do so; nor would Laruelle wish there to be one. This is simple arbitrariness from the perspective of philosophy, but it is an invitation into an experimental and unshackled arena of thought when viewed from the perspective of nonphilosophy.

For Laruelle, the real is both transcendental, albeit immanently, and thoroughly materialized: it is anterior, irreducible, and indifferent to thought, being, and world, but is also at the same time their "real" cause (James 2012, 169). As Laruelle puts it: "Transcendental immanence is the real cause of transcendence," or alternatively, transcendental immanence is the "condition or real base of the sphere of effectivity in general" (1989, 72, 87; 2013b, 71, 85). These formulations touch on one of the other key axioms relating to Laruelle's understanding or experience of the immanent real: that it is determining in-the-last-instance of all transcendence, that is, of all that can appear or be made manifest of thought, being, and world. Laruelle's axiom of determination-in-the-last-instance will be discussed at much greater length in chapter 3.

What this means is that when Laruelle uses the term "transcendental," its sense is different from the sense it has in, say, Kant's transcendental unity of apperception or in the Kantian or Husserlian understanding of the transcendental ego. It has been transformed into the immanent real or One, which is the condition, or more properly speaking the cause or determination-in-the-last-instance, of all phenomenal and worldly transcendence. This also means that when Laruelle speaks of the One as an unreflected, nonthetic, or internal transcendental experience, or as a lived immanence prior to all representation, this is not an experience that is lived as the content of an ego, subject, or consciousness, phenomenological or otherwise. This experience of transcendental immanence is arguably no less strange and paradoxical than Nancy's experience of freedom, and it will be explored and discussed further in chapter 3. Yet in making immanent the transcendental in this way, Laruelle abandons that moment, so decisive within Kantian and post-Kantian philosophies of finitude, that seeks to establish, police, or in some way experience the limits of thought. As he puts it in the context of his more recent nonstandard thinking, the gesture of thought at stake here is one that "abandons the touching of edges that philosophy engages in" (2008, 77). In this way, Laruellian nonphilosophy is both vanishingly close to the Nancean image of philosophy and simultaneously decisively distinct.

Ēpimētheia and Plasticity: Stiegler, Malabou, and the Origin of Philosophy

If Nancy's thinking of philosophy at its limit is intimately bound up with an experience of "the origin removed from every logic of origin," and if Laruelle's nonphilosophical axiomatics of immanence affirms a "autonomous and original milieu of existence without transcendent content" (Laruelle 2012, 306), then it is certain that the images of philosophy offered in the work of Bernard Stiegler and Catherine Malabou are no less preoccupied with the question of origin and its withdrawal from thought. Stiegler and Malabou are closer to Nancy insofar as they present a Derridean- and Heideggerian-inflected image of philosophy in the wake of deconstruction (as opposed to Laruelle's "non-Heideggerian deconstruction"; Laruelle 1989, 179–212; 2013b, 177–210). However, their respective accounts of the origin of philosophy differ significantly. Stiegler adopts a narrative relating to philosophy's historical inception (its diachronic origin, if you like), which then offers a template for interpreting different philosophies while dictating how philosophical thinking itself should be practiced. Malabou, on the other hand, offers a structural (more synchronic) account of philosophy in its originary or primordial moment that describes how and why philosophy is subject to diachronic change as such.

The Forgetting of the Origin

Any reader of Stiegler's work will know that his characterization of the Western philosophical tradition as the forgetting of technics informs the core of his practice as a philosopher; more specifically, it informs the central arguments of his first major work, *Technics and Time 1: The Fault of Epimetheus* (1994, 1998). Given that this characterization is both central and well known, its presentation here will be schematic and oriented specifically toward the way the question of origin in Stiegler is treated in chapter 4.

In *Technics and Time 1* Stiegler presents readings of the Greek myth of Epimetheus and of ancient Greek philosophy in order to argue that "at its very origin and through to the present moment, philosophy has repressed technics as an object of thought. Technics is the *unthought*" (1994, 11; 1998, ix). Arguably, this claim provides both the inaugural moment and the guiding impetus for all of his subsequent thinking throughout the 1990s through to the present day.

The repression or forgetting of technics, or its status as the unthought of thought, is located in the separation Stiegler discerns within ancient philosophy between knowledge itself and the supposedly instrumental technicity of language and of technical objects more generally: "At the beginning of its history philosophy separates *tekhnè* from *epistémè*, a distinction that had not yet been made in Homeric times" (1994, 15; 1998, 1). In its moment of origin, Stiegler argues, Western philosophy instantiates a hierarchy among knowledge (of being, of nature, of essences) and the instruments, techniques, or technical objects that articulate knowledge (of measurement, calculation, inscription, and recording). According to this hierarchy, technics (*tekhnè*) plays no role in the constitution of knowledge (*epistémè*) but rather is purely instrumental; it is merely a neutral vehicle for the articulation of knowledge as such. This, for Stiegler, is exemplified most clearly in Aristotle's *Physics*, in which, he argues, the theory of causes does not endow technical objects with any causal power of any kind: "No form of 'self-causality' animates technical beings. Owing to this ontology, the analysis of technics is made in terms of ends and means, which implies necessarily that no dynamic proper belongs to technical beings" (1994, 15–16; 1998, 1). In opposition to this understanding of technicity as instrumental, all of Stiegler's complex readings of bioanthropological, phenomenological, and postphenomenological discourses in *Technics and Time 1* (e.g., Bertrand Gille, André Leroi-Gourhan, Heidegger, Husserl,

Gilbert Simondon, and Derrida) are orientated toward a determination of technicity as constitutive—constitutive of time, memory, history, and the human per se. It is in this context that the separation and hierarchization of *tekhnè* and *epistémè* at the origin of philosophy is understood as a repression or forgetting of the former that was effectuated as a means of affirming the autonomy, self-presence and grounded or unmediated status of the latter.

In the myth of Epimetheus, this forgetting of *tekhnè* at the origin of philosophy is read by Stiegler as philosophy's forgetting of the originary lack of origin within the human in general and of technics as a supplement to the default of the origin. This myth is of course well known, and Stiegler's use of it has been well documented (Lewis in Howells and Moore 2013, 53–68). Michael Lewis has usefully noted that the names of Epimetheus and Prometheus map exactly onto philosophical-anthropological insights relating to originary lack and to technical supplementation as constitutive of the human. So, for instance, natural science and anthropology have told us much about the distinctly human inability to "survive without an extended period of technical or prosthetic support" (Howells and Moore 2013, 64); that is to say, infant humans require a sustained period of care from adult humans before they can survive alone, and adult humans require the most basic forms of technicity (relating to the provision and preparation of food, such as hunting, farming, and cooking, and of shelter, such as clothing and habitation) in order to survive at all. Lewis also points out that Stiegler cannot rely solely on the absorption of scientific, anthropological, and ultimately purely empirical information into philosophy in order to sustain his argument. This is because the argument in favor of "the co-originarity of man and technics"—drawing as it does on broadly post-Kantian, (post)phenomenological philosophical contexts (Husserl, Heidegger, Derrida) and on historical or bioanthropological accounts of humans and technicity (Gille, Leroi-Gourhan)—"amounts to a mutual contamination of the empirical and the transcendental," and it does so in such a way that "the empirico-transcendental contamination that defines the human automatically generates a mythical retelling of its own origin" (Howells and Moore 2013, 60, 64–65). If the originary co-constitution of the human and technics, understood as a constitutive lack and/or default of the origin within the human as such, were narrated solely as an empirical-historical sequence of events, such a narrative would itself constitute an origin, a grounded and purely scientific-empirical explanation of the human. It is just such a ground or origin that is lacking according to

Stiegler's account, and therefore the recourse to myth becomes a mode by which originary disorigination can be adequately described and in which originary "empirico-transcendental contamination" can be inscribed. The story of Epimetheus's forgetfulness as he fails to allocate any attributes or capacities to humans, and of Prometheus's subsequent theft of the *tekhnè* of fire, thus become an image that, as myth or as a kind of phantasm of the origin, gives a narrative of originary and empirico-transcendental lack without itself describing a purely empirical or historical-chronological moment of origin. It describes a quasi-transcendental moment (to use Derrida's term) in which lack is primordial and, as Stiegler puts it, "the inscription of elementary supplementarity in the empirical is originary" (2001, 254). This originary co-constitution of the human and the technical describes human thought and consciousness as an emergent and entirely material dimension that has worldly technicity as its condition of possibility. This is what Stiegler will come to call organological conditioning, and insofar as he will also come more explicitly to understand technicity as an extension of material life itself (as discussed in chapter 4), the organological conditioning of human consciousness at the same time represents a naturalization of the transcendental moment within thought in general.

In the light of Stiegler's use of the myth of Epimetheus, the image of philosophy as a forgetting of technics at the moment of its historical origin ultimately emerges as an image of philosophy inaugurating itself in a repression of a more originary empirico-transcendental or quasi-transcendental forgetting, lack, or default of the origin that constitutes the human as such. From this point on, the entirety of his philosophical project is to both locate instances of the forgetting or repression of originary technics within the thought of individual philosophers and to diagnose the symptoms of the Western tradition of the philosophical forgetting of technics within wider cultural and historical becoming. So, for instance, even those philosophers whom he relies on most and who have given extensive accounts of technicity and the role of "calculating technique" within modernity—thinkers such as Husserl, Heidegger, Simondon, and Derrida—all are diagnosed as having key blind spots or moments of insufficient analysis with respect to the originarily constitutive role played by technicity. In relation to all these thinkers, and to others Stiegler draws on as positive resources for thought, his key conceptual moves are to highlight the useful ways in which they problematize the constitutive role of technology and technicity, and then point out that they never go quite far enough.

What these thinkers fail to do is to fully and comprehensively approach the origin of philosophy, its originary forgetting. They fail to encounter the outer limit, or limitation of philosophy, as it is marked by the loss or default of the origin that constitutes the human in general, and human thought and consciousness in particular. Stiegler argues of philosophy, as the questioning of being and as a truth or meaning of being that is always, in its first moment, a quest for beginnings, that "philosophy is essentially, at least at its beginning, a search for the origin" (2003, 21; 2009, 7). If he is correct that the inception of the human is a lack of foundation that is articulated in its co-constitution with the supplementarity of technics, then it is not surprising that in its foundationalist or metaphysical mode (that is to say, the Western tradition), philosophy will seek to repress or forget this originary lack. The only means philosophy can therefore have to overcome metaphysics is to encounter and uncover the lack or loss *of* origin and *at* the origin, and thereby also encounter its own limits—that is to say, the constitutive limitation that is set by originary lack and supplementarity on its foundational ambition as philosophy. In this sense, for Stiegler, the act of postmetaphysical philosophizing becomes an experience of limits and of a loss of origin or foundation within both thought and being. This brings his image of philosophy vanishingly close to Nancy's experience of freedom, limits, and ontological groundlessness as discussed earlier: "The passage to the act of philosophy would have a relation to limits, a radical experience of limits of which the first name [. . .] would be 'origin'" (2003, 24; 2009, 9).

Yet if for Stiegler the practice of philosophy is always engaged in an experience of the limits of philosophy itself, carried out by way of an uncovering of the originary lack and supplementarity of technics, this does not restrict his thinking to a series of more or less deconstructive commentaries on canonical thinkers in which the repression of technics is brought to light in each. Perhaps the most consequential imperative that derives from Stiegler's image of philosophy is the demand that the forgetting of technics be addressed as such within wider industrial and postindustrial culture and historical development. Philosophy's instrumentalization of technicity and its downgrading to a secondary, subordinate, and nonconstitutive position has wide-reaching consequences, which are also the task of philosophy to interrogate: "Whereas, on the one hand, the understanding of technics is now, as it has been since the industrial revolution and the profound social changes that accompanied it, largely determined by the categories of ends and means, on the other hand, technics has itself achieved a new opacity, which will be more and more difficult to explicate" (1994, 28; 1998, 14). The uncovering of the (default of) the origin of philosophy and the originary lack of the human dictate that philosophy begins to explore the constitutive role played by technics and to uncover, analyze, and respond to what emerges in this act of uncovering. All of Stiegler's key themes and concerns are articulated here. These include individual and collective individuation, the development of the culture industries in the twentieth and twenty-first centuries, the synchronizing of libidinal desire and the massification of consciousness in late capitalism, the pharmacological status of technics, the analysis of contemporary digital culture, and its social and political implications.

Crucially, in the context of this discussion, Stiegler's uncovering of originary technics is concerned with the question of knowledge, with the constitution and transmission of knowledge in and through different modes of technicity (instrumentation, writing, digital and information technology). To uncover the constitutive role played by originary technics is to pose the question of technologically mediated access to knowledge: "We raise this question at a moment when, in the perspective of contemporary technics, the technologies of the elaboration, conservation, and transmission of forms of knowledge are undergoing radical transformation, profoundly affecting the order of knowledge. But what is knowledge as such if it is transformable in this way? Can one say it has a 'unity'?" (1994, 217–18; 1998, 210–11). In this context, the image of philosophy that Stiegler presents also, as it does with Laruelle, implies a thorough rethinking of the conditions of knowledge and a concomitant questioning of the distribution and interrelation of different modes of knowledge. At the same time Stiegler presents an image of philosophy that, as it does in Nancy, engages an experience of ontological groundlessness and of the limits of philosophy itself. It is in this context that the relation of philosophy to science in Stiegler's thinking will be understood and a specifically Stieglerian naturalism elaborated.

The Metabolism of the Origin

In *Plasticity at the Dusk of Writing* (2010), a synoptic overview of her personal development as a philosopher, Catherine Malabou offers a general account of philosophy itself. The story she tells is one in which her own philosophical perspective develops out of a reading or emergent understanding of the history of philosophy and what this says about its deep structure and the conditions of its transformation and change over time. She informs her readers that from the beginning of her career, from the earliest work on plasticity in Hegel through to her analyses of change and metamorphosis in Heidegger's thinking of Being, "the history of philosophy appeared to me less as a single history than as a cleavage between two histories, two conceptions of history and two conceptions of philosophy" (2005b, 21–22; 2010, 7). The insight she develops is essentially drawn from Derridean deconstruction, and it echoes Nancy's formulations relating to the foundational ambition of metaphysics that he developed in his earliest commentaries in the 1970s. According to this account, traditional metaphysics, in the very moment that it lays philosophical foundations, also at the same time encounters an absence of ground and an impossibility of foundation. In this context, Malabou indicates that she came to discern within philosophy itself a "sharing of *traditional* philosophy and its 'destruction,'" adding immediately, "I soon accepted [. . .] that from this point on any philosophical doctrine would necessarily be worked through and fragmented by its own 'destruction,' which is its *paradoxical contemporary*, and that all the temporal differences at work within a single thought function a posteriori, even though originally, from the dislocating force of the metaphysics that we have not yet finished interrogating" (2005b, 22; 2010, 7). Yet whereas Nancy in the 1970s derived his understanding of the simultaneous grounding and ungrounding of foundationalist metaphysics from deconstructive readings of thinkers such as Kant and Descartes, Malabou narrates the manner in which her own understanding of the constitutive "dislocating force of metaphysics" emerged from her detailed engagements with Hegel and Heidegger (James 2012, 85–95). There is a divergence in Malabou's interpretation of the tradition of philosophy. Where Nancy traces a historical trajectory according to which the philosophical decision of Western thought reaches a certain limit point in Kant's architectonic of pure reason, and where Stiegler traces an inaugural forgetting of technics within philosophy that is more or less maintained until he himself fully and thoroughly uncovers it as such, Malabou argues that all philosophy has always already been divided by the double sharing of foundational gesture on the one hand and the destruction, dislocation, or withdrawal of foundation on the other, "as if philosophy and the end of philosophy took place *simultaneously*" (2005b, 22; 2010, 7).

What this means is that the equivocation and non-self-identity of ontological and metaphysical foundations in philosophy occur, as it were, structurally and necessarily in and of themselves; they do not require a deconstructive uncovering, activity, or perspective on the part of the philosopher to make them effective as such. The equivocation and non-self-identity of philosophical foundations has always and already been at work anyway as the condition both of philosophy's production and of its mutation or transformation over time, or as Malabou herself puts it: "Destruction (*Destruktion* or *Abbau*) is not the consequence of a methodological decision of the thinker but rather an internal, immanent movement of philosophical content" (2005b, 43; 2010, 18). This is a radicalization of one of deconstruction's central insights, and it is central to the argument relating to plasticity in Hegelian thought that Malabou develops in her first major work, a doctoral thesis written under Derrida's direction. In this book, the plasticity that was shown to inform the work of the Hegelian dialectic is viewed as a key motor and structure of philosophy: "In *The Future of Hegel*, plasticity already designates the ability of the dialectic—and beyond it all traditional philosophy—to *negotiate with its destruction*" (2005b, 57; 2010, 27).

It is in this context that the motif or schema of plasticity becomes the center point for Malabou's thinking as a philosopher and for her conception of philosophy in general. Plasticity for Malabou is the condition of the development and transformation of philosophical concepts and discourses over time; it is "the *metamorphic structure* that authorizes the shift from one era of thought and history to another" (2005b, 57; 2010, 27). At the same time plasticity is also "the *metabolism of philosophy*, the exchanges arranged between its inside and outside, itself and its other" (2005b, 57; 2010, 27). This deep structure of philosophy is one according to which the forms and formed concepts or discourses that constitute it are always exposed to a constitutive alterity, excess, and futurity (the "future" described in *The Future of Hegel* [1996, 2005a]), which means that they can never be fixed or stable, that they will always be inhabited by an exteriority that will undermine their self-identity and be the condition of their transformation into future philosophical forms, concepts, and discourses. This exposure to alterity and excess is also an exposure to an instance of ontological void or groundlessness, and it offers the key to the way in which scientific and posttranscendental or postphenomenological arguments are linked in Malabou's thinking.

It is worth noting the extent to which the plastic "metabolism" or "metamorphic structure" of philosophy in Malabou's work emerges across her analyses of Hegel and Heidegger as a function of the plasticity and metamorphic structure of being itself. This specifically feeds into her understanding of the diverse images of being that philosophy produces as fantasmatic, phantasmagoric, or imaginary (2004a, 95; 2011, 71; James 2012, 93–94). The schema of plasticity presents an image of philosophy as a series of forms, concepts, or fundamental images of being that are necessarily exposed to transformation and change, but this also and necessarily means that being itself is always and only presented or thought by way of a plastic image. In this way, the schema of plasticity and the metamorphic structure of philosophy it dictates "allow us to define the object of philosophy in a radically new manner: as an *imaginary object*. This imaginary 'object' is being itself, the powerfully hallucinatory effect of its phenomenon" (Malabou 2005b, 77; 2010, 39 [translation modified]). In this way, the ever-transforming and changing forms of philosophy function as a manifestation of a wider economy of change and exchange of being as such, as it is imaged in philosophy but also, and crucially, as it is made manifest in the mutability and plasticity of the material forms of existence more generally. This notion of thought as a material form situated within a wider economy of material existence is central to my discussion of Malabou's understanding of cerebral plasticity and epigenesis in chapter 4.

One might wonder, however, whether Malabou is not engaged in self-contradiction or inconsistency. It is arguable that the image of philosophy as plastic or metamorphic and the image of "being" in general as fantasmatic and imaginary offer a kind of metaimage of both that is not itself mutable but rather invariant or static—an image of plasticity and change that governs everything, such as philosophy, being, and the forms of material existence, without itself being subject to the logic of transformation it describes. Malabou is aware of this objection and takes pains to point out that what is at stake in the "image of philosophy" she offers is not in fact a paradigm as such, nor a static or invariant code (as it is, say, in Laruelle), but rather an understanding or experience of form that arises in the wake of deconstruction:

> The idea of a "structure of philosophy" does not therefore refer to a paradigm, model, or invariable; rather, it describes the result of the destruction of the paradigm, model, or invariable in general. *By "structure of philosophy," I mean the form of philosophy after its destruction and deconstruction. This means that structure is not a starting point here but rather an outcome.*

> *Structure is the order and organization of philosophy once the concepts of order and organization have themselves been deconstructed. In other words, the structure of philosophy is metamorphosized metaphysics.* (2005b, 97–98; 2010, 51)

The metamorphosis and transformation of form (that is to say, plasticity) is something that occurs and is recognized as such in an experience of thought and philosophy as they unfold or are produced in the aftermath of metaphysical foundationalism. Plasticity thus offers a general schema of or transcendental condition for thought, and with this articulates an image of philosophy as such. Yet it does so only as a kind of self-awareness of thought itself as it immanently experiences its own absence of foundations, and with this the metabolism of its originary mutability. The naturalization of the transcendental condition of thought in Malabou, understood as plasticity, is discussed more fully in chapter 4 in relation to the paradigm of Malabou's epigenesis.

Just as Nancy and Laruelle in their different ways have been shown to engage in a practice of thought that emerges in an experience of thought itself—one that unfolds as radical suspension of philosophy's power to secure an origin, ground, or foundation—so too do Stiegler and Malabou root their thinking and philosophical practice in an experience of the (absence of) origin. The experience of the forgetting or default of the origin (Stiegler), or the metamorphic or mutable metabolism of the origin (Malabou) means that, as with the experience of limits in Nancy or of radical immanence in Laruelle, thought itself is not grounded or founded as such. Rather, it emerges as a movement, opening, or production of (non)philosophical practice in the absence of foundation or ground. As such, the emergence or movement of (non)philosophical thought can be understood not just as a specific mode of experience (of limits, immanence, of the default or plasticity of the origin) but also as a mode of experimental technique.

Thought as Experimental Technique

In order to draw out further the stakes and implications of this image of philosophy as an experimental technique, and more specifically in order to begin to question how the relation of philosophy to science might be framed in this context, it is worth bringing it into comparative contrast with one of the best known images of philosophy that has emerged in recent decades within the context of the Continental tradition. This of course is the definition of philosophy offered by Deleuze and Guattari in their late collaboration, *What Is Philosophy?* (1991, 1994).

The definition they give is well known: philosophy, Deleuze and Guattari argue, is "the art of forming, inventing, of fabricating concepts," or precisely and more rigorously still, it "is the discipline that involves creating concepts" (1991, 8, 10; 1994, 2, 5). As will become clear, this understanding of philosophy as the creation of concepts differs markedly from the images of philosophy that emerge in the work of Nancy, Laruelle, Stiegler, and Malabou. Perhaps most obviously, Deleuze and Guattari have no interest in any affirmation of the death of metaphysics or of the end or overcoming of philosophy as such; indeed, they are highly dismissive of what they would no doubt think of as an all too Heideggerian thematic (1991, 14; 1994, 9). What this means is that they, unlike, say, Nancy and Stiegler, do not conceive of philosophy as involving or engaging an experience of limits. It also means that, unlike all the four thinkers treated here, they do not problematize the question of metaphysical foundations in a deconstructive manner and therefore do not explicitly engage with an experience of an absence or void of foundation in terms of an impossibility or limitation of philosophical conceptuality. In this context, a sharp difference emerges between the Deleuzo-Guattarian understanding of philosophy as the creation of concepts and the practice of thought as experimental technique as it is being elaborated here. In the light of this decisive difference, the specificity and novelty of the relation between philosophy and science in Nancy, Laruelle, Stiegler, and Malabou can begin to be discerned, and in this way, the outlines of a post-Continental naturalism can emerge.

In Deleuze and Guattari, the relation between philosophy and science is clearly delineated and their difference from each other sharply demarcated. Where philosophy is the creation of concepts on the plane of immanence, science is the determination of functions on a plane of reference. Deleuze and Guattari make it plain that concepts and their creation are the exclusive domain of philosophy but that this in no way implies that philosophy itself has some kind of exclusive privilege or preeminence in relation to science or to other areas of knowledge, activity, or creativity (a claim that, as will become clear, I contest) (1991, 14; 1994, 8). In this way, and to a certain extent, they echo Heidegger in *Was heißt Denken?* when he affirms that science does not think because, they insist,

"It is pointless to say there are concepts in science" (1991, 37; 1994, 33). The plane of immanence on which philosophical concepts are created is not, it should be remembered, itself a concept; rather, it is the image of thought as such, the image thought gives of itself to itself in order to orientate itself as thought (1991, 41; 1994, 37). The plane of reference in relation to which scientific functions are determined specifies science as the domain of knowledge that "concerns itself only with states of affairs and their conditions" (1991, 38; 1994, 33).

What this means is that thought, understood as the, or a, (virtual) plane of immanence, knows no limitation and is not limited in the way that (actual) states of things or their conditions can be said to be limited. The creation of concepts in Deleuze and Guattari always occurs in an infinity of the virtual: "The creation of concepts has no other limit than the plane they happen to populate; but the plane itself is limitless" (1991, 79; 1994, 78). As the determination of functions on a plane of reference, science renounces the state of limitlessness, or in-finitude, that characterizes the virtuality of thought. It does so because it binds itself to the finitude of actual states and things (1991, 118; 1994, 118). This affirmation of the immanence of thought, its boundless potentiality, and the concomitant eschewal of any experience of, or play at, the limits of thought is something that Deleuze and Guattari share with Laruelle. In this regard, it is notable that they explicitly acknowledge Laruelle as pursuing one of the most interesting endeavors of contemporary philosophy (1991, 45n5; 1994, 234n5). Arguably, though, and in the absence of Laruelle's axiomatics of radical immanence and of the unknowability of the real, such an eschewal of any experience of the limit leaves Deleuze and Guattari open to the criticism that, despite their protests to the contrary, they are nevertheless privileging the discourse of philosophy, maintaining its mastery over other forms of discourse, and preserving its traditional power and sufficiency to conceptually determine being and the real.

This absence of engagement with the limits of thought can be discerned in the contradictory ways that Deleuze and Guattari characterize the plane of immanence itself. They are clear on the one hand that all philosophies will sketch out different planes of immanence, their respective images of thought, upon or within which they will create the concepts that make them into distinctive bodies or schools of philosophy. Yet on the other hand, and ultimately, all philosophies are situated within the plane of immanence as such, the virtual domain of thought, which can never be known or given a determinate image: "We will say *the* plane of immanence is, at the same time, that which must be thought and that which cannot be thought. It is the nonthought within thought. It is the base of all planes, immanent to every thinkable plane that does not succeed in thinking it. It is the most intimate within thought yet the absolute outside—an outside more outside than any external world because it is an inside deeper than any internal world: it is immanence" (1991, 61; 1994, 59). This reference to the plane of immanence as that which must but which also cannot be thought, and to the language of absolute exteriority (which is also an interiority), is surprising to the extent that it is not accompanied by any thinking of the limit, which nevertheless and paradoxically forms the ungraspable horizon of thought's limitlessness. The footnote to Blanchot's *The Infinite Conversation* (1969, 1993) that accompanies this passage is instructive in this regard. In Blanchot, as in Nancy, the experience of radical exteriority, of that which is in excess of thought as an absolute unknown, is always also a paradoxical experience of the limit, or it is the "limit-experience" (*expérience-limite*) of which Blanchot (1969, 300–342; 1993, 202–29) speaks in relation to Bataille's thinking. As Blanchot puts it, "The limit-experience" is "the inaccessible, the unknown itself," and if thought is indeed infinite, an experience without limits (as Deleuze and Guattari, like Blanchot, affirm), then it is only because there is a "limit-experience," that paradoxical moment within thought at which "perhaps, the limits fall but that reaches us only at the limit" (Blanchot 1969, 305, 311; 1993, 205, 210). So Deleuze and Guattari invoke or affirm a radical nonthought, or an unthinkable and unknowable exteriority within thought, and they do so with direct reference to Blanchot. Yet at the same time they deny that thought can occur as a limit-experience because it unfolds on and as a plane of immanence that is infinite and without limit.

In the absence of an experience of the limit of thought or of a limit-experience like that articulated by Blanchot (and similarly in Nancy), it can be questioned whether Deleuze and Guattari can ever adequately encounter or account for the radically unknowable as such. They nevertheless insist that such an encounter should be the preeminent task of philosophy, its "supreme act" being to show that the plane of immanence is "there, unthought in every plane" (1991, 62; 1994, 59). Yet if philosophy's actual task is to create concepts simply by way of giving consistency to a unlimited and infinite virtuality that allows for no experience of the limit of the knowable or the thinkable, then it is difficult to see how the process of concept creation takes the radically unknowable into account or experiences it in any way at all and as such. Concepts are creatively produced and fashioned on a variably configured, but in each case limitless, plane of immanence, but they never engage an experience of absolute exteriority or the radically unknowable. This would be an essentially Blanchottian, Nancean, and perhaps also Derridean critique of Deleuze and Guattari's understanding of philosophy as the creation of concepts.

Such a critique can be supplemented from a Laruellian perspective that, affirming radical immanence and abandoning any play with the limits of thought as it does, might at first sight be seen as closer or more sympathetic to the Deleuzo-Guattarian point of view. However, for Laruelle, the immanence thought by Deleuze and Guattari is never thought radically enough. Their practice of philosophy remains all too philosophical insofar as its thinking of immanence continues to be constituted in a mixing or synthesis of immanence and transcendence, of interiority and exteriority. Deleuze and Guattari's attachment to and faith in the operativity of philosophical concepts mean that they maintain philosophy's power of representing the real in the concepts they themselves create (difference, virtuality and actuality, the univocity of being, and so on) and in so doing preserve its traditional authority, mastery, and the structural violence of the philosophical decision as such. This at least would be Laruelle's critique of Deleuze and Guattari's philosophical practice, one that is developed at length in relation to Deleuze's "idealist version of difference," in particular in *Philosophies of Difference* (1986, 188; 2010b, 170).

Deleuze and Guattari's image of philosophy as the creation of concepts on the/a plane of immanence thus posits a radical unthought within thought, affirms an absolute exteriority as thought's most intimate interiority, and thereby seeks to affirm also "the inaccessible, the unknown itself" (Blanchot 1969, 305; 1993, 205). Yet in eschewing the possibility of anything like a Blanchottian or Nancean limit-experience within thought, and in the absence of a nonphilosophical or Laruellian critique of philosophical conceptuality as such, philosophy, as understood and practiced by Deleuze and Guattari, signally fails to encounter or really account for "the inaccessible, the unknown itself."

This failure is arguably borne out in the way in which Deleuze and Guattari configure the relation of philosophy to science. In conferring upon each their distinctive, separate, or opposed domains (concept creation, function determination), they arguably maintain the autosufficiency of each, their self-positioning and complementarity: "It could be said that science and philosophy take opposed paths, because philosophical concepts have events for consistency whereas scientific functions have states of affairs or mixtures for reference" (1991, 127; 1994, 126). This complementarity of the production of events on the one hand and the reference to states of things on the other is, nominally at least, underpinned by Deleuze and Guattari's avowal of the unthought or the unknowable within thought. Philosophy and science both contain or include "an *I do not know* that has become positive and creative, the condition of creation itself, creation that consists in determining *by* what one does not know" (1991, 129; 1994 128). Yet if one accepts the argument that the eschewal of an experience of limits and/or the absence of a nonphilosophical (Laruellian) axiomatics in Deleuze and Guattari's thinking means that they do not encounter or adequately account for the unknown as such and thereby remain on the side of the already determined and determinable, then their accompanying account of the creativity of both philosophy and science is at best called into question.

It is notable and significant in this context that those thinkers and scholars who have explored the relation of philosophy to science in the wake of *What Is Philosophy?* (1991, 1994) have focused on the complementarity between Deleuze and Guattari's philosophical concepts on the one hand and scientific theories as they are already established, known, and constituted on the other. John Protevi has pointed out that Deleuze himself makes it clear that his aim is to provide a metaphysics for contemporary science. Protevi also points out that a good number of scholars have sympathetically treated the relation of philosophy to science in this context by relating Deleuzian ontology to dynamic systems modeling (2013, 1; see also Ansell-Pearson 1997, 1999; Bonta and Protevi 2004; de Bestegui 2004). Exemplary in this regard is the work of Manuel DeLanda. DeLanda's groundbreaking *Intensive Science and Virtual Philosophy* (2002) puts Deleuzian and Deleuzo-Guattarian thought into relation with contemporary geometry, complexity theory, and chaos theory. Like Deleuze and Guattari themselves, DeLanda underlines both the independence of contemporary scientific theories and "virtual philosophy" from each other, but also their complementarity insofar as they ultimately describe the structure of the real in a similar manner. This leads DeLanda to conclude that "if the same conclusions can be reached from entirely different points of departure and following entirely different paths, the validity of those conclusions is thereby strengthened" (2). This implies a relation of reciprocity between philosophy and science according to which the distinct authority and autosufficiency of each is mutually reinforcing as they come together in shared conclusions relating to the determined and determinable truth of being or of the real. Despite DeLanda's argument that both (analytical) philosophy and science emerge transformed into novel configurations in the encounter between the two that he stages (6), the image of authority and autosufficiency that he attributes to both is arguably traditional or familiar insofar as it does not really take into account the radically unknown. The encounter that is staged is one between known philosophical (Deleuzo-Guattarian concepts) and established, albeit contemporary, scientific theories.

It is in this context that the image of philosophy as an experimental technique such as it has been elaborated here in relation to Nancy, Laruelle, Stiegler, and Malabou emerges as entirely different from the image of philosophy as the creation of concepts. Where Deleuze wishes to maintain philosophy as a metaphysics that would be adequate to contemporary scientific knowledge, all the four thinkers treated here seek to overturn, suspend, or otherwise move beyond philosophy's metaphysical foundationalism in a variably configured experience of an absence of origins and foundations. To this extent, they also sharply differentiate themselves from, and innovate in relation to, the Francophone Deleuze and Guattari–inspired attempts to renegotiate the relation between philosophy and science. In *Cosmopolitics 1*, for instance, Isabelle Stengers invokes the necessity of a speculative operation of thought understood as a "thought experiment" (2003, 20; 2010, 12). She does so in a manner that appears at first to resonate closely with the notion of thought as experimental technique elaborated here (and with the "speculative" as it is understood in relation to Nancy in chapter 2). It is clear, however, that for Stengers, this experience of thought is to be understood within the context of a Deleuzo-Guattarian metaphysics of the virtual and the actual, and according to their distinctive image of philosophy as a creation of concepts. The role of speculative or experimental thought here is that of "creating possibles" and of "creating words that are meaningful only when they bring about their own reinvention" (2003, 20; 2010, 12–13). This Deleuzian metaphysics of creativity underpins or informs much of Stengers's attempts to innovate in relation to both science—for example, in *Cosmopolitics 1* in the context of thermodynamics (2003, 175; 2010, 179)—and philosophy—as elsewhere in the context of her reworking of the thought of Alfred North Whitehead in *Thinking with Whitehead: A Free and Wild Creation of Concepts* (2002, 287–310; 2011, 275–76). Stengers does at times invoke the notion of a "passage to the limit" in her account of Whitehead (2002, 31, 296–97; 2011, 263). She also does this, for instance, in relation to Bruno Latour's thought in *The Invention of Modern Science* (1993, 173–74; 2000, 153). In the case of Latour, though, this passage to the limit relates to the limit of a specific received concept (that of a "parliament"), which is reconfigured as the "parliament of things" in a rhizomatic structure that, true to the Deleuzo-Guattarian metaphysics of immanence, would be "without limits" (1993, 74; 2000, 153). In the case of Whitehead, the encounter with the limit is a matter of exceeding traditional existing forms of rationalism in order to liberate thought for the creation of concepts (2002, 296–97; 2011, 262–63). In neither case is it a question of encountering the limit of thought as such according to the specific logic of thinking at the limit that Nancy elaborates in *The Experience of Freedom* (1988, 1993b). For all their originality, therefore, Stengers's reworking of the philosophy–science relation and her innovative, constructivist account of scientific knowledge and understanding both remain firmly within the orbit of a Deleuzo-Guattarian metaphysics.

In Deleuze and Guattari, I have argued, philosophy and science come together in a complementarity of philosophical concept creation on the one hand and the determination of functions on the other. In so doing, they reciprocally reinforce their traditional self-sufficiency and authority as discourses of truth and as (metaphysically grounded) determinations of the real. They reproduce the same, or a similar, relation of mutual benefit and reciprocal bestowal of authority that was discerned in the continuity between philosophy and science discussed in the context of American naturalism in the introduction.[4]

In contrast, and as will be demonstrated throughout the discussions of the following chapters, in the thought of Nancy, Laruelle, Stiegler, and Malabou, philosophy and science come together; they find their point of jointure or suture in their shared experience of limits (or in the case of Laruelle of a single "frontier") and of an absence of metaphysical or ontological ground or foundation. This experience of limits is different from the Kantian limitation of knowledge via the distinction of the phenomenal and the noumenal. It is an experience in which thought, encountering the impossibility of grasping its conditions of possibility, is exposed to the always excessive singular plurality of existence to which it is nevertheless always already in some kind of relation.[5] The continuity between philosophy and science posited in the naturalism of Sellars, Quine, and Lewis appears to be broken insofar as philosophy here has been shown to derive images of itself without any reference to the authority of scientific knowledge but rather in relation to an autonomous experience of thought derived from engagements with philosophy itself. Arguably that margin of autonomy, independence, and regulative or speculative activity that was identified in the work of the American naturalists (and is arguably an irreducible dimension of any philosophy) is broadened and given a primacy here that makes the traditional naturalist philosophy–science continuum questionable. Yet as nonfoundationalist and nonphilosophical, and as articulations of thought at the limit, there is no return here to a first philosophy that is abandoned by naturalism in favor of scientific knowledge. As experimental, thought here is unmoored from the pretensions of traditional first philosophy and free to, as it were, experimentally engage with the sciences anew. Yet most decisively, and as will become clear in the following chapters, both philosophy and science are no less encounters with and of the real. They have their cause or determination in the real and therefore articulate a radical realism of both philosophical thought and scientific knowledge.

While they are both reconfigured in such a way as to reduce their traditional metaphysical and epistemological sufficiency in certain specific ways, philosophy and science also gain by dint of being articulated in new, arguably more real, relations to the real—relations that have jettisoned the illusions of philosophical and metaphysical foundationalism. The task will be to show that in the new relation of philosophy to science that these different techniques of thought articulate, the outcome is not a relativism or any kind of a constructivism. Rather, the outcome is an entirely novel realism and with this a reconfigured, thoroughgoing, and nonreductive naturalism. The preceding discussion has begun this task by discerning new images of philosophy that collapse or void the transcendental moment of thought in a fourfold manner: into the fundamental orders of ungrounded transimmanence and of radical immanence (Nancy and Laruelle), and into the primordial yet disoriginary conditioning of technicity or plasticity (Stiegler and Malabou). The elaboration of post-Continental naturalism can now proceed from these renewed experiences of thought to a radical rearticulation of the continuity between philosophy and science that has been one of the core defining attributes of both analytic and Continental naturalism to date.

2 THE RELATIONAL UNIVERSE

NANCY'S PURELY FORMAL, MINIMALIST ONTOLOGY of the singular plural describes above all a relational universe and a world in which objects and entities are constituted in and through the relation of singularities. The passage from such an ontology, derived as it is from a critique of Kantian reason, to a naturalistic philosophy aligned with scientific knowledge may seem to be far from self-evident. Yet the experience of thought exposed at the limit of philosophical reason, as outlined in the previous chapter, leads Nancy in just such a direction.

Nancy's is a minimalist ontology and derives from an experience of both thought and existence as a groundless singular plurality. If such an ontology really does describe all that can be said of being as being (albeit minimally, experimentally, and speculatively), then it should describe the being of the natural world and of all the singular entities within it, and it should describe also our own place as humans amid that natural world and among other nonhuman entities. This is to say nothing more than that any ontology, however experimental or speculative, should in principle be able to account for the independent existence of the natural world as well as our place within it, and thereby be brought into some kind of relation with the natural sciences which also take that natural world as their object of knowledge. As will become clear, Nancy's limit-thought and the minimalist ontology to which it gives rise allows for an understanding of nature and of the cosmos in which the relation between philosophy and science is not one of a seamless continuity exactly, and even less one of subordination. Rather, both philosophy and science, working at their limits and at the limits of what current scientific knowledge tells us about the universe, engage with each other in a more open-ended and less hierarchical manner.

In *The Sense of the World* (1993c, 1997) Nancy evokes the possibility of renewing the old category of "philosophy of nature" within contemporary thought (64, 40). He does not, however, fully develop this idea. Arguably, however, Nancy's thinking of sense, world, and cosmos in this seminal work offers something like a protophilosophy of nature that today, over twenty years later, can be more explicitly developed as such. A number of contemporary philosophical contexts provide an important backdrop for such a task and serve also to highlight the specificity, originality, and interest of Nancean philosophy and its possible relation to science. First, one might cite attempts within philosophy of science in recent years to develop a fully naturalized metaphysics that sets itself up in contradistinction to and against analytic metaphysics. Most prominent in this regard would be James Ladyman and Don Ross's 2007 polemic, *Every Thing Must Go*. Second, it would be impossible not to cite the development within Continental philosophy, also since 2007, of speculative realism and its critique of correlationism within the broader post-Kantian trajectory of European philosophy. As a movement, speculative realism is associated with four key figures: Graham Harman, Ray Brassier, Quentin Meillassoux, and Ian Hamilton Grant. However, their respective bodies of thought are in fact diverse, leading to questions as to whether speculative realism really refers to a philosophical movement as such (rather than, say, to a more diffuse tendency). Third, the recent attempt by Unger and Smolin to develop a natural philosophy that would go under the name of "temporal naturalism" provides both an interesting contrast and a partial ally to the natural and cosmological thought that can be developed from *The Sense of the World*. Each of these contexts will be engaged with in some detail in what follows. More immediately, however, it might be noted that they each offer different accounts of the relation of philosophy to science and imply different perspectives on the possible "naturalization" of thought and philosophy.

The project of naturalizing metaphysics undertaken by Ladyman and Ross, and others is generally and overtly scientistic in its ambition. Ladyman and Ross begin *Every Thing Must Go* with an open defense of scientism proposing "a metaphysics that is motivated exclusively by attempts to unify hypotheses and theories that are taken seriously by contemporary science" (2007, 1). Not only will this naturalized metaphysics take scientific knowledge as its exclusive and ultimate authority, but it will also restrict its scope to the activity of unifying science as a whole in order to produce a coherent scientific world picture that brings together in consilience fundamental physics and all the other special sciences. The target of Ladyman and Ross's polemic is the tradition of analytic metaphysics that, they argue, proceeds by arguments not informed by scientific knowledge because it is "based on prioritizing armchair intuitions about the nature of the universe over scientific discoveries" (2007, 10). In a sweepingly dismissive and questionable polemic, Ladyman and Ross argue that analytic metaphysics, insofar as it does not sufficiently ground itself in science and the scientific image of the world, "fails to qualify as part of the objective pursuit of truth, and should be discontinued" (2007, vii). Similar views are repeated in the essays collected in *Scientific Metaphysics*, edited by Ladyman and Ross with Harold Kincaid (2013). Here it is argued unequivocally that "speculative ontology has no place in an objective scientific understanding of the world," that naturalized metaphysics must be "inspired and constrained by our best science," and that therefore philosophy is "continuous with science," to the point, one might argue, of being almost entirely absorbed by it and subordinated to it (22, 33, 40). As will become clear, such radical scientism is potentially problematic insofar as it tends to be rapaciously eliminative and as it denies the objective or fundamental existence of large swaths of experience. These include, unsurprisingly, the denial of traditional metaphysical notions of purpose in nature, the meaning of life, free will, and historical teleology. Yet more questionably it includes also the elimination of the fundamental existence of entities or things, of qualia, and even of thought or consciousness itself.[1]

In contrast to this attempt within contemporary philosophy of science to create a fully naturalized metaphysics, speculative realism does not offer a uniform position with regard to the relation of philosophy to science. Graham Harman has acknowledged general agreement that it can be separated into two more or less opposing groups.[2] These two groups he calls epistemist and antiepistemist, with Quentin Meillassoux and Ray Brassier being placed in the former and himself and Iain Hamilton Grant in the latter (2012, 23). Without going into too much detail, what Harman's distinction rightly identifies is a difference of orientation with respect to mathematical and scientific forms of knowledge within speculative realism. On the one hand, Meillassoux endorses a kind of mathematism when he claims that mathematics is able to describe those aspects of reality that are human independent (2010, 26). On the other hand, Brassier endorses a version of scientism, inspired by his interest in the thought of Wilfrid Sellars, when in *Nihil Unbound* (2007, 16) and elsewhere he privileges the scientific image of the world.[3] In both cases, mathematical or scientific thought are privileged insofar as they offer the possibility moving beyond or of stripping away the finitude of human knowledge in order to approach and attain the reality of the world as it would be without humans. Harman and Grant conversely, in their antiepistemic speculative-realist approach, hold (in quite different ways) that ultimate, human-independent reality is withdrawn more absolutely from all thought or phenomenal appearance. So Harman, for instance, develops his philosophical theses using a mode of argumentation, by turns phenomenological and metaphysical, which is entirely divorced from scientific knowledge. As Peter Wolfendale notes, Harman bases his arguments on "a special kind of intuition unknown to the sciences," his thinking being "independent of the sciences insofar as it is based on a form of evidence that is entirely alien to them" (2014, 104–5). Arguably, therefore, speculative realism, if it is indeed a movement, can be separated into one strand that aligns itself with science and a second strand that unambiguously separates itself from, and maintains an autonomy with regard to, scientific knowledge.

In contrast to both contemporary scientific or naturalized metaphysics and speculative realism, the natural philosophy proposed in Unger and Smolin's *The Singular Universe and the Reality of Time* (2015) situates itself squarely within the domain of current and possible future, scientific debate, and practice while explicitly avoiding scientism. Their temporal naturalism argues that much of modern day scientific theory, and in particular theory within the physical sciences, is still beholden to rationalist metaphysical prejudices and to a metaphysically informed overestimation of the ontological status of mathematics. What scientific theories do, they argue, is allow us to give powerful mathematical descriptions of subsystems of the universe, but do so at the expense of downgrading or even eliminating the temporal-historical becoming of the cosmos as a whole. By assuming that the reality of the universe is wholly isomorphic with the mathematical objects used to describe it, and by not accounting for the extent to which mathematics may fail to describe the flux and dynamism of temporal becoming, Unger and Smolin argue that current science has locked itself into a series of impasses, impossibilities, and paradoxes that can be overcome only if the universe is viewed as a singular whole that is historical through and through and that is evolving according to a fundamental order of time. What is interesting about Unger and Smolin's project, pursued as a collaboration between a philosopher and a cosmologist, is that it does not allow thought to defer to or humble itself before science.

Their proposed philosophy is, no doubt controversially, extremely critical of the fundamental metaphysical disposition of many modern physical theories while in no way disputing their specific domains of validity when they have received ample empirical verification. At the same time, their temporal naturalism is thoroughly informed by current scientific knowledge, with its attainments as well as its gaps and impasses, while allowing itself to move speculatively beyond the limits of that knowledge both to make general philosophical arguments and to propose a specific agenda for the future of empirical research within fundamental physics and cosmology. To this extent, Unger and Smolin's project represents a more mobile and less hierarchical relationship between science and philosophy than is found in either the scientism of contemporary naturalized metaphysics or that of the epistemic speculative realists Meillassoux and Brassier.

Nancy's singular plural ontology is not in any way derived from science. It does not seek to gain any authority, vindication, or validation from scientific knowledge. It is elaborated from a critique of Kantian rationality and from a specific experience of thought. To dismiss such an ontology out of hand because it is broadly speculative and does not take or seek any authority from the sciences implies that they alone give us privileged access to an understanding of reality and that other ways of knowing are somehow deficient or lacking. Such a scientistic attitude is itself a specific philosophical position that needs to be defended as such and that is never a simple or straightforward given. However, this is not to say that an ontology such as Nancy's cannot be placed into a relation with scientific knowledge and be found wanting or lacking. Fundamental ontological positions that appear to fly in the face of, or be flatly contradicted by, scientific theories should be discarded or questioned, or at the very least enter into a sustained negotiation with such theories and both the empirically verified facts and metaphysical assumptions that inform them. At the same time, philosophy, as a thinking *of* and *at* the limits of thought, may have a valuable role in thinking *at* or even *beyond* the current, or indeed constitutive, limits of scientific knowledge in a manner that can guide future research, both experimental and theoretical, or otherwise allow for a better understanding of those limits.

If there is to be a passage from Nancy's singular plural ontology to a more fully developed naturalistic philosophy, then such an ontology needs to be placed into a critical and comparative relation with scientific thought and understanding. Such a relation will not be one of subordination but rather one in which both philosophical and scientific thought, exposed to and at their limits, maintain a less hierarchical and more open stance to each other. Both share a tendency to view the being of the world in purely relational terms. In what follows, this relational understanding of the world, shared by both Nancean ontology and by some modern scientific theories, will be explored in the context of three distinct areas of questioning.[4] First, the question of the status and fundamental existence of things and discrete objects or entities will be interrogated. Nancy's account of the "sense" of the world will be developed in relation to debates within contemporary scientific metaphysics and speculative realism. Second, the problem of life and of the individuality of the living organism will be addressed in relation to Georges Canguilhem's philosophy of biology and more recent thinking about life that has emerged from cutting-edge work in biochemistry. Nancy's philosophy of sense will be further developed here on the basis of parallels that can be drawn between it, Canguilhem's biological thought, and contemporary empirical research into the question of life. Third and finally, the question of the structured unity, nonsubstantial being, and dynamic becoming of the cosmos will be posed. Nancy's collaboration with the theoretical physicist and cosmologist Aurélien Barrau is of key importance here, as is the project of temporal naturalism currently being pursued by Unger and Smolin. Out of this shared image of a relational universe and this threefold questioning of things, life, and cosmos, a unique and distinctive naturalism can be elaborated from Nancy's minimalist singular plural ontology.

Things

Two fundamental insights inform Nancy's ontology, first that being is relational and must therefore always be understood prepositionally in terms of "being-to" (*être-à*), and second that such prepositional being can be given the name of "sense" (*sens*). These two insights give rise to a vision of the world (understood in its most inclusive sense as that which surrounds us in the human and natural world as well as in the wider cosmos or universe) in which objects or things are constituted purely in relations of singular existences but gain their autonomy or individuality as things by dint of the "sense" of their being, or more precisely by dint of their "being" being understood as "sense." As will become clear, the understanding of existence as relational is strongly correlated with modern scientific thought, and in particular with the scientific image of the world in the wake of quantum theory. Understanding being as "sense," however, finds no such obvious or immediate correlation with fundamental physics and should be seen as a purely speculative move on Nancy's part that gives a name to being as such in a manner that science does not or cannot do. Nancy's use of the term *sens* is complex. In the original French, the word designates meaningfulness, sensory perception, and direction. Perhaps most importantly, Nancy's naming of existence as "sense" articulates his understanding of being in general as foundationless or without ground and as in excess of anything that might take the name of "substance."

These two insights of singular plural ontology, affirming being as relational and naming it as sense, place Nancy's thinking in a middle ground between the scientific metaphysics of Ladyman and Ross and the object-oriented speculative realism of Graham Harman. Ladyman and Ross's self-declared scientism leads them to argue for a structural-realist position that eliminates objects or things from the most fundamental level of existence or objective being. Harman's object-oriented ontology, on the other hand, affirms that the structures and relations grasped by scientific knowledge are subtended by a deeper, hidden layer of reality in which objects have an autonomous and substantial existence inaccessible to science and its experimental methods. Where for Ladyman and Ross relations are primordial and objects are to be eliminated from ontology, for Harman it is objects that are ontologically fundamental and relations that are always secondary, derived, or, he argues, vicarious. Nancy's thinking aligns itself with that of Ladyman and Ross insofar as his vision is relational and structural, but it distances itself from them insofar as he understands objects or things to be constituted out of relations in such a way that they nevertheless still have some kind of ontological status or autonomy as distinct and singular entities. This is inadmissible for Harman, who holds that the real existence of objects or things can only be secured on the basis of affirming their being as autonomous substances, withdrawn from all relation and being-to.

Nancy's understanding of being as sense is derived from the same experience of thought that leads him to affirm being as singular plurality. With the withdrawal of a substantive ground for being, or in the absence of some principle that would confer upon being unity, substance, and plenitude, the question of the overarching meaning, value, or purpose of existence is displaced or modified. We can no longer ask if existence has any meaning and seek to discover such meaning in a unitary figure or principle that would in some way lie outside of existence—for example, in a world of timeless essences or of mathematical and logical form, of the image of a creator God, of a providential historical process or a rational telos that would underpin temporal becoming. Yet at the same time the world around us appears to be abundantly meaningful and replete with the possibility of intelligibility and sense constantly offered up to the senses. This leads Nancy to argue that "the world *no longer has* a sense, but it *is* sense" (1993c, 19; 1997, 8). Sense is not sought in a unitary principle outside of existence but inheres plurally within or more precisely as existence; or, as Nancy puts it, "The exigency of sense is nothing other than existence insofar as it *has* no sense" (1993c, 20; 1997, 9). Sense, therefore, is the potentiality of relational existence or being to yield intelligible structured organization or meaningfulness, and this potentiality constitutes being as such. All that is, is or makes sense, and it does so in the absence of any unifying ground, each and every time singularly, and always in a network of relation to a plurality of other singular existences. To the extent that there is anything, it simply *is* as sense, and that is the most fundamental thing that can be said of its being: "The *there is* makes sense by itself and as such" (1993c, 18; 1997, 7).

The emphasis here on the "there is" (*il y a*) of existing things making sense "by itself and as such" is absolutely decisive because it underlines that Nancy's ontology is not restricting itself to thinking being or sense such as it relates to humans or human apprehension. Rather, sense understood *as* being describes or names the existence of the "there is" *as it is*, in and of itself and in excess of any correlation with human signification or phenomenal consciousness. In Heideggerian terms, it names being as it is independently of the being-there of Dasein. On this point Nancy could not be more explicit: "There is something, there are some things, there is some there is—and that itself makes sense, and moreover nothing else does. It does not make sense only for, through, or in Dasein" (1993c, 92; 1997, 55). This emphasis on the human independence of sense is absolutely decisive because it shows clearly the extent to which Nancy in 1993 is attempting to think outside what Quentin Meillassoux later in *After Finitude* (2010) comes to term "correlationism."

The use of the correlationist thesis to describe the broad trajectory of post-Kantian thought is nowadays well known in the context of speculative realist thought. Opposition to correlationism is perhaps the sole philosophical position shared by all the names associated with speculative realism as such, and according to its proponents, it clearly marks out the novelty and innovation of speculative realism as a philosophical movement. Correlation, Meillassoux argues, is the idea according to which "we only ever have access to the correlation between thinking and being, and never to either term considered apart from the other" (2010, 5). The correlationist thesis effectively distills Kant's Copernican revolution—that is, the claim that we can only know the world through our own human transcendental a priori intuitions and categories—and reads the subsequent philosophical tradition that comes out of Kant as always, in some form, affirming this essentially Kantian claim. The correlationist, according to this account, will necessarily assert that we will never know the thing in itself. We will know only the correlation between ourselves and the thing or between thought and being, and this knowledge, as a correlation, gives us only the identity of the relation, with the subjective and the objective being now forever interdependent and in a form of equivalence each with the other (5–6).

Yet Nancy is clear in his desire to move beyond a correlationist position by naming being, albeit speculatively and minimally, as sense. Insofar as there are things in the world, the "there is" of those things makes or "is" sense in and by itself and as such, and not just in correlation to human conscious or to the being-there of Dasein. The full extent and explicitness of Nancy's noncorrelationist ontology of sense is made clear:

> The world beyond humanity—animals, plants, and stones, oceans, atmospheres, sidereal spaces and bodies—is quite a bit more than the phenomenal correlative of a human taking-in-hand, taking-into-account, or taking-care-of: it is the effective exteriority without which the very disposition of or to sense would not make . . . any sense. One could say that this world beyond humanity is the effective exteriority *of humanity itself*, if the formula is understood in such a way as to avoid construing the relation between humanity and the world as a relation between subject and object. For it is a question of understanding the world not as man's object or field of action, but as the spatial totality of the sense of existence, a totality that is itself *existent*, even if not in the mode of *Dasein*. (1993c, 92; 1997, 55–56)

Understanding the natural world, "beyond humanity" as an "effective exteriority" of things which *are*, and therefore *make*, sense is to say nothing other than that such things already *were*, and *made*, sense in themselves long before human consciousness or being evolved in that world. It is also to say that they will continue to do so when humans have become extinct. Understanding the natural world in this way is to say that humanity, such as it is, is never centered in on itself or cut off from nature; rather, it is always constituted in and through the exteriority of nature itself in a structure of relations that make up the totality of the space of natural existence understood as sense.

There is a strong realism underpinning Nancy's thinking of the sense of the world. Sense here exists prior to human perceptual consciousness and in excess of our systems of symbolic meaning or signification. Nancy is trying to think a "sense that precedes all senses, and that precedes us, warning and surprising us at once" (1993c, 11; 1997, 2). To this extent, the ontology of the sense of the world needs to be thought outside of the orbit of any phenomenology because, just as the speculative realists will later come to argue, phenomenology is centered on phenomenal appearance as it is manifest to human consciousness.[5] Sense is in excess of phenomenal disclosure; or more precisely it is an absolute exteriority with regard to phenomenal appearance. Nancy puts this in the following terms: "Phenomenology does not open us up to that which—in sense and consequently in the world—infinitely precedes consciousness and the signifying appropriation of sense, that is, to that which precedes and surprises the phenomenon in the phenomenon itself, its coming or its coming up" (1993c, 32; 1997, 17). Because phenomenology remains tied to appearances as they are manifest for human consciousness, the only way for thought to open itself to that which infinitely precedes consciousness and human signification is to expose itself, at constitutive limits, to that which is in excess of thought and thereby shift onto the terrain of the speculative. This is precisely what Nancy's thinking of the world and sense does.

Yet the understanding of being as sense, in excess of human consciousness and therefore entirely independent of it, needs to be reconciled with Nancy's vision of being as relational, as always a being-to. As was indicated earlier Nancy's use of the term *sens* in French exploits the polyvalence of its meaning. *Sens* refers to meaningfulness per se, the fact that something has a disposition toward intelligibility. It refers also to sensation or the sensible contact of one thing as it touches another. Last, *sens* can mean direction or way (as in the French *sens unique* for one-way direction of travel). Exploiting all these three meanings, Nancy argues that the "there is" of things in the world *is* sense, in and of itself, but is so only insofar as sense is always also a plurality of instances relating to other instances and in that "relation to" forming the extension or space of the world as such. It is in this sense that, for Nancy, "*world* means at least *being-to* or *being-toward* [*être-à*]; it means rapport, relation, address, sending, donation, presentation *to*—if only of entities or existents to each other" (1993c, 18;

1997, 8). Things have their being as sense only insofar as they are constituted as such in their relations with one another, in the relational movement of the sense of things to each other, and in their differentiation from each other in just such a movement. This relationality of being-to is prior to or in excess of phenomenal manifestation (of things to humans, but also of nonhuman things or entities to each other) while at the same time being the condition of possibility for the emergence of phenomenal or sensible-intelligible appearance as such.

So Nancy's is a realist vision of a structured and relational world (of sense), in which human being is just one moment within the wider structure of relations in which the being of anything can only be thought in terms of its constitution in and by way of the exteriority of the relational whole. In this context, the world outside of man, the human-independent world of real things and objects, is also always a surrounding world of nature from which we as humans are never ontologically separated. Such a surrounding world of nature will precede and succeed us, but we have emerged as humans only in relation to it and will become extinct as humans, always still a relational part of it. This realist vision, insofar as it incorporates human being into the natural world of relations that constitutes it, is a novel form of naturalism.

Information and Sense: Must Every Thing Go?

This realist vision of a structured, relational world is both similar to and different from that presented by Ladyman and Ross in *Every Thing Must Go* (2007). As the title of their book suggests, one of its key arguments is that objects, entities, or individuals have no existence as such at a fundamental ontological level. The mathematics of fundamental physics, most notably that of quantum mechanics (QM), and in particular quantum field theory and phenomena such as entanglement, tells us that at the microphysical level, it makes no sense to talk about little objects or things. In the context of their self-declaredly scientistic approach, the injunction to eliminate things from fundamental ontology is absolute. Scientific knowledge wields total authority here. As Ladyman and Ross put it: "Science respects no domain restrictions and will admit no epistemological rivals [. . .]. With respect to anything that is a putative fact about the world, scientific institutional processes are absolutely and exclusively authoritative" (28). The epistemological authority of science translates directly into ontological authority for naturalized metaphysics and does so according to what Ladyman and Ross call the primacy of physics constraint, or PPC. The PPC articulates the demand that all the hypotheses within the special sciences should be in conformity with the current consensual knowledge regarding fundamental physics (i.e., the quantum theory of today) (44). In turn, a naturalized metaphysics that promotes a credible scientific realism will also be underpinned by fundamental physics and will not deviate from it.

Within this constraint, Ladyman and Ross adopt and defend the doctrine of ontic structural realism (OSR), their own specific version of this doctrine being also known as information theoretic structural realism (ITSR). On the basis of the insights of quantum theory, this doctrine is eliminativist with regard to self-subsistent individuals and holds that "relational structure is ontologically fundamental" (2007, 130) and that therefore "individuals are nothing over and above the nexus of relations in which they stand" (138). Such an eliminativism needs to be viewed, argue Ladyman and Ross, in terms of "the positive thesis that the world is structure and relations" (153). There are objects in their metaphysics, of course; it is simply that "they have been purged of their intrinsic natures, identity, and individuality, and they are not metaphysically fundamental" (131). This OSR is also information theoretical insofar as the concept of information is the fundamental category according to which physical reality is known or determined. Citing physicists such as Jeffrey Bub, Anton Zeilinger, and John Archibald Wheeler, Ladyman and Ross argue that information may come to be "regarded as primitive in physics" and "may in some sense be all there is," insofar as it is "impossible to distinguish operationally in any way reality and information" (186, 186, 189). This leads them to conclude therefore that "all things physical are information theoretic in origin" (212).[6]

Given their disregard for the "armchair intuitions" of analytic metaphysics and their call for research in this area to be discontinued, one can only imagine what Ladyman and Ross would make of speculative ontologies emerging from the Continental tradition of philosophy and derived from postmetaphysical deconstructions of canonical texts. Yet they nevertheless share with Nancy an understanding of the world as composed of structure and relations and of individual entities or things being constituted only in such structures and relations. It is undoubtedly true that there can be no direct or immediate equivalence between the scientific concept of information and Nancy's use of the term "sense." The scientific concept of information has widespread use in both fundamental physics and biology, derived from Shannon's information theory (Shannon 1948; Shannon and Weaver 1949), and Ladyman and Ross specifically indicate that their own use of the term derives from the Shannon–Weaver model of communication (2007, 220). Yet what the use of the concept of information within their structural-realist ontology shares with Nancy's own ontological use of the term "sense" is a capacity to speak or give a name to the relationality of existence that does not imply or conceal any thinking of substance. As Ladyman and Ross note (citing philosopher of physics Christopher Timpson), information is an abstract mass noun that "does not refer to a spatio-temporal particular, to an entity or a substance" (185).[7] That information is not a substance or thing is absolutely crucial here. Both information and Nancean sense offer, in physical science and speculative ontology, respectively, a way of talking about reality in the absence of substantive ground or foundation. In a more recent text, Ladyman and Ross are explicit on the question of the absence of any all-embracing ground that would provide a foundation for being, and for the being of things in particular. The information theoretic conception of existence "is compatible," they argue, "with the possibility that there is no, general, over-arching account of the relationships" that make up all that is (Ladyman, Ross, and Kincaid 2013, 121). They also explicitly deny that there is any convincing basis at all for believing that a "ground of things" exists as such (136).[8]

Therefore, Nancy's ontology of sense and OSR/ITSR both share, albeit in different ways, a strongly realist understanding of existence where things are composed out of networks or relational structures in the absence of any underlying substance. Yet where the reliance on the concept of information leads Ladyman and Ross to be eliminativist with regard to the fundamental existence of things, Nancy's ontology of sense allows their being to still be affirmed on a fundamental ontological level. The full extent of the eliminative ambition of naturalized metaphysics can be seen in Ladyman and Ross's use of the term "real patterns." Borrowing and modifying the term "real pattern" from Daniel Dennett, they argue that "to be is to be a real pattern" (2007, 253). Therefore, to mathematically or probabilistically determine a real pattern according to the physical information that can be known of that pattern "is to say everything there is to be said about the criteria for existence" (121). This leads us to an extreme eliminativism whereby fundamental ontological reality is granted only to that which can be accessed by scientific and mathematical formalization of physical structural relations. This is therefore a scientism that disqualifies from the realm of ultimate real existence not just individuals and things but also the qualia of phenomenal and sensory experience as well as any dimension of existence or experience not reducible to, or formalizable and verifiable by, scientific procedures that determine the information borne by real patterns. Ladyman and Ross's scientism cannot account for, or confer the reality of real being upon, those dimensions of existence—for example, perceptual, affective, or symbolic—that are irreducible to such scientific determination.

The scientism of contemporary naturalized metaphysics generally tends toward an uncompromisingly eliminativist attitude. This may seem highly problematic to those working outside the sciences, in, say, the humanities or areas of the social sciences, who investigate phenomena that appear to have a real and not just epiphenomenal existence yet that are not amenable to the experimental methods of scientific inquiry. So for instance Ladyman and Ross approvingly cite Peter Unger's view that qualia are "the unknowable non-structural components of reality" and affirm that, in their view, "qualia are idle wheels in metaphysics and the PPC imposes a moratorium on such purely speculative philosophical toys" (2007, 154). Yet our immediate perceptual consciousness tells us that qualia, understood in the most general sense as the phenomenal character or qualities of experience, do indeed exist in an indubitable fashion even if that existence is uniquely "for us" and subjective or intersubjective. That science cannot measure or quantify, and therefore know, such qualitative experience does not necessarily offer sufficient grounds for denying its existence as such. What will become clear is that Nancy's ontology, in contrast to OSR/ITSR, is noneliminativist with regard to things but also with regard to the qualitative sensible experiences that living things feel or perceive. Yet Nancy's ontology can nevertheless still be aligned with scientific thought without succumbing to the scientism of Ladyman and Ross, and of contemporary naturalized metaphysics.

A key aspect of Nancy's relational vision of the world is that relationality itself produces differentiation and articulates structure as a series of distinct differentials and therefore as necessarily multiple or heterogeneous. In this context, then, understanding being as sense rather than purely as abstract information allows for the determination of matter as difference and of individual things or entities as instances of material difference. Speaking in terms of material difference rather than of information crucially allows the spatiotemporal particularity (Nancy would say singularity) of discrete instances of the differential structure of existence to be maintained. This allows him to speak explicitly of "the *reality* of the *res* as material difference" (1993c, 96; 1997, 57). Of matter itself, he says, "Matter means here: the reality of the difference—and différance—that is necessary in order for *there to be something and some things* and not merely the identity of pure inherence [. . .]. Matter is a matter of real difference, the difference of the *res*: if there is something, there are several things: otherwise there is nothing, no 'there is.' Reality is the reality *of the several things* there are; reality is necessarily a numerous reality" (1993c, 96; 1997, 57–58). Whereas physical science talks of matter in terms of mass and energy, as well as perhaps ultimately, as has been suggested, in terms of information, Nancy focuses on the formal quality of matter as differentiation across the temporal spacing or extension of relational structure and the way that such differentiation necessarily produces multiplicity. He resists the notion that matter can be understood as some kind of uniform impenetrable density or thickness immanent to the form of things as they are "in-themselves" ("a pure inherence"), for this would no doubt simply be another way of talking about it in terms of substance. For Nancy, the difference or differentiation of matter is that by which "*something* is possible, as *thing* and as *some*: that is, other than as the indistinct inherence or hardening of a one that would not be *some one*" (1993c, 95; 1997, 57). It is by insisting on such differentiation that Nancy avoids eliminativism with regard to things at the fundamental ontological level. Whereas the fundamental physics privileged by Ladyman and Ross's PPC speaks only in the abstract in terms of purely informational real patterns, Nancean sense speaks of a relational existence whose formal articulation as difference maintains the spatiotemporal singularity and distinctness of individual things as they are constituted in their very relationality.

Relations without Relata

One of the most common objections brought to bear on purely relational accounts of existence is that they fail to account, in logical terms, for the fact that distinct entities need to exist first before they can be brought into any kind of relation as such. This is an objection Graham Harman makes with regard to both Ladyman and Ross's OSR and Nancy's ontology of sense.

Harman (2010) offers a highly critical response to *Every Thing Must Go*, arguing that ultimately, whether they like it or not, their scientism, for all its attempt to naturalize metaphysics and affirm a structural realism, nevertheless ends up caught within the logic of Kantian correlationism. According to Harman, insofar as Ladyman and Ross maintain that all those nonstructural aspects of existence that are unknowable to science (things-in-themselves, qualia) are "speculative metaphysical toys," they necessarily uphold the essentially Kantian correlationist claim that we can only know being in its relation to us. As Harman himself puts it, according to OSR/ITSR, "Everything boils down to a correlation between physical structure-in-itself and mathematical structure-for-living-creatures" (2010, 784). If all we can know is that which correlates to our mathematical knowledge of the physical universe, then we necessarily and inescapably remain Kantian correlationists. This aspect of Harman's criticism is questionable: although he clearly signals Ladyman and Ross's own explicit rejection of any residual Kantianism in their position, he arguably fails to account for their eliminativsm with regard to what science cannot determine. For OSR/ITSR, there is simply no residual in-itself or noumenal reality residing behind what science determines. This point has been made explicitly by one of Ladyman's collaborators, Steven French, who points out that in OSR/ITSR, "there are no unknowable *objects* lurking in the shadows and objectivity is understood structurally, in terms of the relevant sets of invariants" (Butterfield and Pagonis 1999, 203). Harman does not engage fully enough, perhaps, with the fact that Ladyman and Ross are explicitly proposing a scientific realism in which reality would be known only as structure but also as it is and independent of humans, such knowledge being proposed as the power and the ultimate privilege of the mathematical and information-theoretic determination of reality. It would not, therefore, be a matter of a correlation of physical structure (in-itself) with mathematical structure (for us) as Harman claims, but rather an isomorphism or total identity between physical and mathematical structure. This places Ladyman and Ross ultimately within the realm of mathematical realism and philosophical-mathematical Platonism rather than Kantian correlationism because formal-mathematical objects never exist simply "for us" but have a human-independent reality.

Just as important in this critique is Harman's raising of the, for him, "reasonable objection that there can be no relations without relata" (2010, 786). Specifically he alludes to the "mystery of how a continuum of relational structure without individual zones would differ from the monism of a whole without parts" (787). The reasoning here is clear: if there are only relations, then there can in reality be no distinct things to relate to each other, and therefore there cannot ultimately be any distinction between a purely relational structure and an undifferentiated, monistic whole. One can immediately see here how this criticism of Ladyman and Ross's structural realism would, for Harman, also apply to Nancy's singular plural ontology.

Indeed, Harman (2012) has made exactly this criticism, and in so doing has similarly attempted to recuperate Nancy's thinking into the logic of Kantian correlationism. In responding to Nancy, Harman draws on selected passages from the essays "The Heart of Things" and "Corpus" as they appear in the volume of collected pieces translated into English under the title *The Birth to Presence* (1993a).[9] Harman appreciates that Nancy is interested in bodies and things, and that the ontology of sense therefore has some filiation with philosophers he considers to be his partial allies in "carnal phenomenology," most notably Emmanuel Levinas and Maurice Merleau-Ponty (Harman 2012, 96). Yet his criticism centers on, and takes great issue with, Nancy's argument that "insofar as it is posited, exposed, insofar as it is the thing itself, every thing is whatever" (Nancy 1990, 206; 1993a, 174). Harman takes this to mean that for Nancy, there must be some layer of being that precedes the relational contact of things with each other, and that this preceding layer must be a shapeless, homogeneous, or otherwise uniform mass of being that only comes into determinacy or distinctiveness in and through relational contact. He also takes relational contact to imply representational manifestation and thereby concludes that, for Nancy, "whatever is withheld from representation cannot have a form" (2012, 102). This leads him to conclude further that "for Nancy all determinacy of bodies is accessible, and a whatever that withholds itself from access must be indeterminable" (103). Because Nancy ultimately holds that the most primordial layer of being is an undifferentiated "whatever," it is clear, according to Harman, that his relational ontology is indeed indistinguishable from an ontology that would affirm "the monism of a whole without parts." This also leads Nancy to adopt or be situated within the "basically Kantian position in which things are granted form only when they are shaped by some other entity" (2012, 103). Nancy therefore is both an ontological monist and, whether he likes it or not, a post-Kantian correlationist.

Yet once again it is arguable that this does not do full justice to the positions that he seeks to recuperate within the correlationist circle. First, one might remark that from his earliest attempts to develop an ontology in works such as *The Inoperative Community* (1986, 1991), Nancy has explicitly opposed and attempted to deconstruct any thinking of being understood as a totality or undifferentiated unity of an immanent existence. This ambition also underlies his reading of Kantian reason and freedom in *The Experience of Freedom* (1988, 1993b), opening the way for all his mature and later works. Second, it might be reiterated that Nancy's relationality of sense is situated prior to phenomenal manifestation and representation. It is therefore not correct to claim, as Harman does, that Nancy "makes the fateful assumption that whatever is withheld from representation cannot have a *form*" (Harman 2012, 102) because for Nancy the differentiation of matter occurs in a relational structure that exists in and of itself prior to any representation. This is to say that the relation of being-to of one thing with another is not yet representation but some kind of contact or touch that precedes representation as its condition of possibility.

It seems that Nancy's statement that "every thing is whatever" means something quite different from what Harman takes it to mean. Even in "The Heart of Things," from which Harman extensively cites, Nancy is explicit that the "whatever" of things constitutes their most proper affirmation as things, and that it "takes nothing away from the differences between things" (1990, 205; 1993a, 174). In *The Sense of the World,* he indicates that, in this context, a thing is "whatever" because "every one is just as singular as every other one. In a sense, they are indefinitely substitutable, each for all the others, in-different and anonymous" (1993c, 117–18; 1997, 72). This is perhaps an obscure formulation, but what is being said here is that each singular point in a relational structure, each thing or entity, is essentially without essence, that is to say, without autonomous identity and therefore ontologically anonymous. That is to say also that it is without identitarian essence, substance, or substantial being. Whatever identity a singular entity or thing does have, it has relationally in its being-to other things. This is not to say that it exists originally in and as some prior undifferentiated mass; rather, the existence of a singular point in a structure is finite: it is born and dies in anonymity and without identity, but it exists or persists as a differentiated, singular, and therefore discrete entity or thing only relationally. It is not that there is an undifferentiated formless thing, as Harman says, that awaits form through the contact it will gain with other things. In *The Sense of the World,* Nancy is explicit on this point: "The 'someone' does not enter into a relation with other 'someones' [. . .]. The relation is contemporaneous with the singularities. 'One' means: some ones *and* some other ones, or some ones *with* other ones" (1993c, 116–17; 1997, 71). Singularity is always necessarily relational to the extent that something can only be singular if it is in relation to something else. It can have no substantive, grounded being in and of itself and is therefore an anonymous "whatever." In its very differentiation, it is without identity, and indeed existence, when taken alone or in isolation. Nancy's use of the term "whatever" is therefore another means by which he affirms the infinitude of being, its absence of ground or an underlying principle that would give it identity and unity. It is another means of refusing the category of substance. In "The Heart of Things," he is explicit on this point: "There is no 'ground' of the 'whatever': the whatever is difference" (1990, 206).[10] As Nancy says in a footnote to Badiou's ontology of inconsistent multiplicity, a principle is needed for that which is, in principle, not thinkable in terms of unity or totality (1990, 220n1; 1993a, 411–12n16). The "whatever" of singular beings or entities is just such a principle.

Clearly Harman is reading Nancy through the lens of his own philosophical preoccupations and in the light of his insistence that entities or things have a discrete autonomous being that is wholly withdrawn from all relation together with his reliance on the category of substance to secure the identity of that autonomous being—substance being precisely the category that Nancy rejects (Harman 2005, 2007). From this perspective, Nancy's ontology ends up affirming a lack of differentiation despite itself—that is to say, Harman is reading Nancy's position against its proposed or explicit affirmation of differentiation. Nevertheless, it is once again arguable that he does not entirely do justice to the detail and specificity of Nancy's position. His critique, after all, appears to be based on a partial reading of just two selected essays from *The Birth to Presence* (1993a) and does not take into account the detail and broader scope of Nancy's singular plural ontology. What this broader thinking of sense, being-to, material differentiation, and singularity demonstrates is that it is not logically or necessarily the case that there can be "no relations without relata." The world—the natural world around us that persists independent of our human concerns—is nothing other than the spacing of singular and plural relational existence. It is "always a differential articulation of singularities that make sense in articulating themselves, along the edges of their articulation" (1993c, 126; 1997, 78). In this way, "sense is coextensive with the confines of the world [. . .]. The world extends to the extremities of sense, absolutely" (1993c, 126; 1997, 78). The material differentiation of sense, or sense as material differentiation, simply is all that there is, in all its irreducible plurality.

The objection that there can be no relations without relata does not seem to hold when one posits the existence of a material differential structure out of whose relations discrete entities or things are individuated as spatiotemporal singularities and as differentials. For Ladyman and Ross, however, the objection raised by Harman regarding the logical priority of relata over relations simply has no relevance in a naturalized metaphysics that takes science as its sole authority and source of legitimacy. Science, and specifically the PPC, tells us that this apparent logical priority of relata over relations can and must be reversed: "A core aspect of the claim that relations are logically prior to relata is that the relata of a given relation always turn out to be relational structures themselves on further analysis" (2007, 155). As one moves from the largest scale of cosmological reality down to that of the natural world around us, to things or living organisms and the organs and cells that compose them, then further down to chemical or biochemical composition, then still further down to fundamental reality at the quantum scale, it is only ever relations and further relations that exist. As far as Ladyman and Ross are concerned, therefore, Harman's view of objects as withdrawn from all relations and as existing autonomously as substances in "obscure cavernous underworlds" is simply disqualified by what science tells us about the relational universe (2007, 179). In the light of the PPC, such a view would, in particular, be disqualified by phenomena such as entanglement at the microphysical level, where the properties of quanta are shared and must be described in relation each to the other even once they have become separated by some considerable distance in space.

What emerges here is a picture according to which Nancy's ontology of sense, as relational and as without substance or ground, can be aligned with the scientific doctrine of OSR but must at the same time be differentiated from it on two decisive counts. First, the thinking of sense is not itself scientific but rather is derived from speculative thought; it refers to being as it is in excess of all phenomenal manifestation and as the condition of possibility of all manifestation per se. This is not a category that Ladyman and Ross would take remotely seriously. No doubt they would dismiss it as an idle speculative toy. Second, Nancean sense, always in excess of phenomenal disclosure, nevertheless refers to a spatiotemporal particularity or singularity of material difference within the relational structure of everything that is. Where ITSR must eliminate things from ontology, Nancy's ontology of sense maintains their real existence as differentiated, relational sense at the most fundamental level.

The similarity between the ontology of sense and OSR/ITSR may not really be as significant as the sharp differences between them and therefore may be insufficient to claim any real or meaningfully close alignment of Nancean thought with that of natural science. From the perspective of philosophers of science (Ladyman and Ross certainly) the understanding of being as relational and without substance or ground may appear to be trivial or without any value for knowledge as such when it emerges speculatively and without reference to the authority of science itself. If any such alignment of Nancean and scientific thought is to be meaningful, and if anything like a philosophy of nature that is based on this alignment is to be possible, then the category of "sense" has itself to begin to make sense for science and become a category that is in some way or another more relevant to scientific thinking. The use of the concept of information in physics does not appear to be reconcilable with the use of the term "sense" in Nancy's ontology. However, where fundamental physics may appear to lead to an impasse in this regard, biology and biological thought may offer a way forward.

Life

The relational understanding of the universe is well established in the French tradition of philosophy of science. For instance, the doctrine of structural realism within that tradition can be traced back to Henri Poincaré's thought as elaborated in works such as *Science and Hypothesis* (Poincaré [1902] 2016).[11] Ladyman and Ross make significant reference to Poincaré, but they make no reference at all to one of his successors, Gaston Bachelard, who in his 1934 work *La Nouvelle esprit scientifique* [The new scientific mind] argued, seventy-three years before Ladyman and Ross, that the new microphysics opened up a perspective of thought whereby "objects only have reality in their relations" and that "it is necessary to renounce the notion of object or thing, at least in the study of the atomic world" (136, 132). However, it is in the work of Bachelard's successor, Georges Canguilhem, that one finds a distinctively biological philosophy that articulates a thoroughgoing relationist position that, on the level of living organisms at least, is not eliminativist with regard to individuals or entities.

A number of commentators have remarked on the centrality of relationality for Canguilhem's account of living organisms. Paul Rabinow has noted, for example, that for Canguilhem, "the basic unit is a living being that exists in shifting relations with its environment" (1994b, 16). Similarly, in his extended work on anthropology and biology in Canguilhem's thought, Guillaume Leblanc has highlighted the extent to which from the perspective of biology: "Individuality is not a being but a relation to an internal milieu and an external milieu" (2002, 31). Likewise, Jean Gayon, one of the leading contemporary French philosophers and historians of science, has argued that for Canguilhem, "individuals should not be conceived of as beings but as relations" (1998, 308). Canguilhem's relational theory of biological individuation is developed in a diffuse way across a range of his writing, most notably in essays collected in *Knowledge of Life* (1965, 2008) and in a later essay dating from 1966, "New Knowledge of Life" (Canguilhem 1994b, 1994d).

Canguilhem, Life, Sense

Canguilhem's thinking in relation to the individuality of living organisms develops in the context of reflections on the history of cell theory and on the relation between organisms and their environment. It is expressed in the following quotation from "The Living and Its Milieu" as it appears in the volume *The Knowledge of Life*:

> From the biological point of view, one must understand that the relationship between the organism and the environment is the same as that between the parts and the whole of an organism. The individuality of the living does not stop at its ectodermic borders any more than it begins at the cell. The biological relationship between the being and its milieu is a functional relationship, and thereby a mobile one; its terms successively changing roles. The cell is a milieu for intracellular elements; it itself lives in an interior milieu, which is sometimes on the scale of the organ and sometimes of the organism; the organism itself lives in a milieu that, in a certain fashion, is to the organism what the organism is to its components. (Canguilhem 1965, 144; 2008, 111)

Here we have a model of biological individuality that does not rely on the idea of a discrete barrier that would function as an absolute line of separation between the interior of the living organism and its exterior environment. Rather, what is described is a structure in which the living thing consists in nothing but relations within relations. There are diverse biochemical processes and their relations to molecular microstructures that in turn form intracellular structures, with such structures then functioning in relation to the cell, those of the cell to the organ, of the organ to the organism, and of the organism to its environment.

It might initially be difficult to see here how the notion of individuality can be sustained if the relation between the organism and its environment is the same as that between its interior parts and its whole, for would not all distinction between inside and outside be lost, and with that the distinctiveness of the biological individual as such? Would this not then confirm Harman's objection that there can be no relations without relata and that a purely relational structure always in fact returns to the monism of an undifferentiated whole without parts? It is here that the notion of biological relation as a functional relation to a surrounding environment or milieu is crucial, however. When the cell relates to its wider surrounding environment of an organ or of the organism as a whole, it does so functioning in specific ways as a cell, taking its organization at that cellular level as a center of reference within that environment. By the same token, when the living organism functions as such within its habitat or milieu, it does so with reference to its organization at the level of the living organism, and its relations to its milieu again take this level as a center of reference.

This idea of relationality as a system of reference to a wider environment that takes specific levels of organization as centers of reference is decisive for Canguilhem's understanding of life and of biological individuality. "To live," he argues, "is to radiate; it is to organize the milieu from and around a center of reference, which cannot itself be referred to without losing its original meaning" (1965, 147; 2008, 113–14). A center of reference, although it is just one level of organization within a purely relational structure, is nevertheless irreducible as such. This is because the system of reference articulated by biological functional relations is always an articulation of a certain kind of meaningfulness. The relation of a specific level of organization (cell, organ, body) to its surrounding environment is at once a relation and a self-relation. It is a relation *to* a milieu only insofar as it is also at the same time a relation that is biologically necessary and therefore meaningful *for* the center of reference being put into relation. Canguilhem expresses this thought in the following terms:

> A center does not resolve into its environment. A living being is not reducible to a crossroads of influences. From this stems the insufficiency of any

> biology that, in complete submission to the spirit of the physicochemical sciences, would seek to eliminate all consideration of sense from its domain. From the biological and psychological point of view, a sense is an appreciation of values in relation to a need. And for the one who experiences and lives it, a need is an irreducible, and thereby absolute, system of reference. (1965, 154; 2008, 120)

A center of reference organizes itself as a center by way of the meaningful functionality of its relations to its surrounding milieu. This means that as a center of reference, it is not simply reducible to either the interior relations that compose it or to the wider system of reference (the biological milieu) to which it relates. What is decisive here is that for the living organism, such functional relations are lived and experienced as values in relation to biological needs, and insofar as they are lived and experienced in this way, they are lived and experienced as "sense." Canguilhem in this decisive moment avoids any reduction of life and of living being to mechanism or to the simple unfolding of biochemical processes. His reference here to physicochemical science is telling: where such science may allow for a reduction to purely chemical and physical processes, biology definitively does not. As Jean Gayon has put it, citing Canguilhem, biologically speaking, "individuals are beings that have needs, and therefore constitute 'an absolute and irreducible reference system' in a given environment [. . .]. The ontology of life must be subordinated to the axiology of life" (1998, 320). That living is always about the value of life for the living organism that has needs and experiences them as such means that living being, as fundamentally axiological, exists always and first and foremost as meaning or sense—or as Canguilhem puts it, "The being of an organism is its sense" (1965, 147; 2008, 113, translation modified).

This alignment of life and of the being of the living organism with sense is taken further in Canguilhem's work of the 1960s and is clearly expressed in the 1966 essay "New Knowledge of Life." Here Canguilhem once again reiterates his decisive insight that "the living thing is, precisely, a center of reference" (1994b, 352). Yet here the idea of biological relationality as a system of reference and of the individual as a center of reference is elaborated in explicitly material terms, with "sense" being understood as the materiality of biological structures and the vital processes they articulate. Here sense becomes more explicitly identified with the organization of living matter as such. As Guillaume Leblanc has noted, where in *Knowledge of Life* Canguilhem derives his notion of life as sense from a posteriori analyses of biological knowledge, in "New Knowledge of Life," he comes to think of sense in terms of an a priori of vital invention immanent to life (Leblanc 2002, 264). To this extent, Canguilhem comes to think of sense as inscribed in matter: "To define life as sense inscribed in matter is to admit the existence of an objective *a priori*, of a properly material, and no longer only formal, *a priori*" (Canguilhem 1994b, 362; 1994d, 317). Thinking of sense as a properly material and objective a priori allows Canguilhem then to bring it into an interpretative relation with genetics and with information theory such as it was being assimilated into biology from the late 1950s and throughout the 1960s.

Here, the being of an organism understood as "sense" is aligned with the understanding of genetic coding within DNA as a form of information. At the same time Canguilhem comes to understand the relationality of the biological organism by way of analogy with cybernetics and information theory. Leblanc highlights that for Canguilhem, the communication of information from within the genome is to be understood analogously with the communication of information within cybernetic systems (2002, 248). Thus, "sense" understood as "relation to" can now also be interpreted as being inscribed materially in, or as simply being, the genetic code of life. This alignment of sense with the genetic code in analogy with the communication of information within cybernetic systems allows Canguilhem to argue that "life has always done without writing, before writing and without any relation to writing, what humanity has sought to do through painting, engraving, writing and printing, namely, to transmit messages" (1994b, 362; 1994d, 317, translation modified). Here, sense, biological material organization, and the notions of information and message transmission are all brought together to articulate Canguilhem's understanding of life. Life becomes something like a writing before writing, but one that is articulated without any relation to human writing. For the biologist working to understand the sense, code, or informational transmission that articulates the relationality of living organisms, it is not a question of imposing meanings on life but rather of uncovering the sense that is inscribed in matter as life. As Canguilhem puts it, "sense is found and not constructed" (362, 317, translation modified).

This assimilation of sense, genetic code, and information theory brings Canguilhem to the point where he can talk of life in terms of a *logos*. As Jean Gayon has noted: "An encoded genetic information can be called a *logos* [. . .] insofar as it is a material, not abstract, principle of definition" (1998, 323). Earlier, in relation to Ladyman and Ross's ITSR and their championing of the PPC, it was noted that there could be no equivalence between Nancy's use of the term "sense" and the use of the Shannon–Weaver derived concept of information within fundamental physics. Canguilhem, however, in his thought of the 1960s argues that the use of the term "information" within biology and in analogy with cybernetic theory can in fact be aligned with his own ontological usage of the term "sense." He has, however, been criticized for his bringing together of the genetic code with the logos of a message understood in information theoretical terms. Henri Atlan (1999), for instance, in the first volume of *Étincelles de hasard* [Sparks of chance], called into question the alignment of genetic coding with information theory on the basis that it ignores the environmental complexity within which the expression of genes takes place and is therefore too reductive and deterministic. This critique has also been made by Jean Mathiot, who argues that the chemical reality of genetic coding and its construction of biological structures is of such a complex spatiotemporal character that it offers "a causal model of differentiation" that is "inaccessible by way of any determinism operating according to the sole 'logos' of a genetic message" (1993, 202). The complexity of gene expression, which involves both the variability of genes themselves and the multiplicity of forms to which they can give rise, cannot be described in terms of "a deterministic model operating in a straight line—informational or other—passing from one order to another" (202). This leads Mathiot to conclude that "the reduction of the gene to the givennes of an initial instance of information is today impossible" (206n15).[12]

Such criticisms would suggest that even in biology, the attempt to bring together the concepts of sense and information, by relating both to the materiality of DNA and the genetic code, is flawed and is not borne out by what the science of complexity has come to tell us about the processes of genetic expression. However, Guillaume Leblanc has responded to such critiques by noting carefully that Canguilhem in "New Knowledge of Life" is not offering a "history of science" text that argues as a matter of historical fact that science has definitively come to align genetic coding with the concept of information as it is used in cybernetics and communication theory. Rather, Canguilhem is offering "a series of philosophical propositions articulated in and around biology" (2002, 251). Biological philosophy here, Leblanc argues, is unfolding as an "art of reading" that interprets biological facts in a certain manner (252). Building on this insight, it can be argued that Canguilhem is not necessarily posing a strict identity between the scientific-theoretic term "information" and the speculative-philosophical term "sense." Rather, through a philosophical interpretation of biology, he may be suggesting that what science names as information and determines as such through its own theorizing and empirical investigations, philosophy in its own ontological domain names as sense. Interpreting the genetic code as "sense inscribed within matter" (Canguilhem 1994b, 362; 1994d, 317) does not alter or subtract from what biological science in its domain does with the term "information." Rather, it could be said that the term "sense" operates as an interpretative supplement, operating on an ontological level, to the term "information," which operates within biology on a theoretical and empirical level. In this way sense, genetic code, and logos are brought together philosophically solely by way of analogy with information theory and cybernetics, and with the purpose of developing the strictly philosophical argument that life needs to be understood as an objective a priori inscription in matter.

The validity of this philosophical interpretation hangs more on the possibility of understanding genetic code as information in some form than it does on the precise analogy with the more deterministic message transmission within cybernetic systems. One can imagine how the informational model of the genetic code could easily accommodate considerations of biological complexity and find other, less linear and deterministic means of describing the communication of genetic information as such. Therefore, if the genetic code can be understood in some way as information, then the means by which genetic information is communicated can come to be understood differently, and by way of other analogies, without necessarily having to reject the philosophical interpretation that Canguilhem develops in relation to sense, genetic code, and information and without discarding his key philosophical argument that life can be understood as sense inscribed in matter.

This is simply to recognize that any critique of Canguilhem's thought, such as those brought forward by Atlan and Mathiot, needs to recognize and respond to that part of his thinking that is a philosophical reading or interpretation of more empirical biological knowledge. At the same time, it is entirely right to hold that philosophical reading to account in the light of subsequent or further developments within biological knowledge, if only to question how it might philosophically respond to such developments. To a large degree, and since the 1970s, biology does appear to have moved beyond the more reductionist and linearly determinist approaches adopted within molecular biology from the 1950s onward, and complexity theory has, of course, increasingly come to inform biological research.[13]

Canguilhem's use in the 1960s of the more deterministic model of cybernetic information theory should be called into question in the light of complexity theory. Yet if the more philosophical part of Canguilhem's argument is separated from its empirical biological core (relating to the functioning and expression of the genetic code), then there is scope to assess the interest and viability of his thinking of life in relation to the current state of biological knowledge.

A full survey of current biological thinking and research concerning the question of life, which is clearly beyond the scope of this discussion, is provided by Gayon (2010). The research and writing of British biochemist Nick Lane offers a useful and contemporary example, however, of the ways in which Canguilhem's thinking might be assessed in the light of current cutting-edge work within biology that is devoted to exploring the genesis, nature, and development of life on earth. In *The Vital Question* (2015), Lane proposes a vision of the birth and evolution of life on earth based on his ongoing research in the origins of life program at University College London. He also argues that contemporary biology needs to lose its exclusive focus on life as the transmission and expression of genetic code and adopt also what he terms a bioenergetic understanding of living organisms and their structural organization.

The question of analogies with cybernetics and communication theory aside, Lane does not dispute the priority and privilege enjoyed by the concept of information within biological science from the mid-twentieth century onward, and, he notes, it remains the case that "biology is now very much about the information concealed in the sequences of proteins and genes" and that "today biology is information [. . .] and life is defined in terms of information transfer" (2015, 7, 22). Rejecting this exclusive focus on information, his central argument turns on the insight that genetic code alone cannot account for the genesis and morphological evolution of cellular life and that the emergence of such life also needs to be accounted for in terms of energy flux and physical constraint (13, 51). Lane develops a theory of the genesis of life within deep-sea alkaline hydrothermal vents that is based on energy exchange through proton gradients. According to this theory, the initial structure of the first and simplest living cells would have emerged from the complex structure and geochemistry of these alkaline hydrothermal vents. He then traces a narrative of the development of the simplest prokaryotic cells through to more developed eukaryotic cells. This narrative takes the constraints of energy and physical structure to be decisive and at least as important as those of the transfer of genetic information and environmentally driven processes of gene selection. His argument then turns to a focus on the importance of the mitochondria (the energy house of cells) and the role that the energetic dimension of multicellular life may have played in the evolution of sexuation and the distinction between germ cells and the soma, and, with that, phenomena such as aging, cell death, and mortality. At the center of all his arguments is the insight that "energy flux promotes the self-organisation of matter" (94).

These contemporary arguments drawn from research in biochemistry may initially seem to be a long way from Canguilhem's theorizing of life as sense in the 1950s and 1960s. Yet what both Canguilhem and Lane share is the recognition that living organic structure is fundamentally sustained in the maintenance of thermodynamic disequilibrium. Living beings take energy from their surrounding environment, convert that energy into organized structure, and maintain that structure for the duration of their lives through further energy consumption, thus always sustaining themselves in a state of thermodynamic disequilibrium with regard to their surrounding environment. In "The Constitution of Physiology as Science," an essay that appears in the collection *Études d'histoire et de philosophie des sciences* (1994a), Canguilhem notes that at the very least it can be said of living beings that they are "systems whose improbable organisation slows down a universal process of evolution towards thermal equilibrium, that is to say towards the most probable state: death" (1994a, 262). Canguilhem also often cites Bichat's famous phrase from *Physiological Researches upon Life and Death* (1809), published originally in French in 1800, that life is the "the group of functions that resist death." The activity of living is precisely the maintenance of biological structure against the threat of dissolution into thermodynamic equilibrium, that is to say, death. Lane's bioenergetic account of life echoes this clearly. If it is "energy flux" that "promotes the self-organisation of matter," thus constituting life, then it is necessarily the case that the energetic activity of life is in the service of the maintenance and prolongation of living self-organized structure. As Lane puts it: "Cell structures are forced into existence by the flux of energy and matter," and so "all living organisms are sustained by far-from-equilibrium conditions in their environment" (2015, 95).

Canguilhem's understanding of biological sense as the "appreciation of values in relation to a need" and of the living organism as a "center of reference" only has meaning within the context of this thermodynamic understanding of life. A living organism has needs only insofar as it is, as

Canguilhem describes, a system whose "improbable organisation slows down a universal process of evolution towards" the "thermal equilibrium" of "death." The living organism is a center of reference that lives or experiences its relations with the environment as its sense only on the basis of the value that this arresting of the evolution toward thermal equilibrium bestows. What is clear, therefore, is that both Canguilhem and Lane share a vision of life as the self-organization of matter, a vision that requires both the coding of genetic information and the thermal disequilibrium of energy flux across biophysical structures to generate and sustain itself. However, although Lane as a biochemist restricts himself to the discussion of the evolution of the structures and mechanisms of biological energetics (such as redox chemistry and ATP synthase in the mitochondria), Canguilhem the philosopher goes further to interpret the activity of biological organization and its functional relations in terms of "sense."

In the light of Lane's contention that biology needs to understand life as energy flux as well as the information transfer of genetic codes, Canguilhem's biological philosophy of life still remains pertinent and able to yield interpretative insights with respect to current biological knowledge. Indeed, what Canguilhem's thought adds to Lane's account is a way of bringing together both the view of life as genetic information transfer and as energy flux into a philosophy of the living organism understood as a relational center of reference and as sense inscribed in material-biological processes. Canguilhem's thought allows for all the biochemical processes of life to be taken into account and interpreted without reducing the phenomenon of life as such to the simple mechanisms of those biochemical processes.

Philosophy of Nature and Naturalized Knowledge

Nancy's ontology of sense and his argument that the existence of things needs to be understood as material difference or as the being-to of sense can be usefully recast in the light of Canguilhem's biological thought. Canguilhem's argument that life is sense inscribed in matter is an argument pertaining to the ontology of life itself and does not, on the face of it, appear incompatible with Nancy's own ontological use of the term "sense." As will become clear, such an alignment comes on condition of rethinking Canguilhem's sense inscribed in matter in terms of Nancean sense exscribed as matter. If the two philosophies can be aligned in this way, it can be argued that scientific thinking can after all make sense of Nancean "sense." Nancy's ontology can indeed describe the natural world as it is known and thought by science—or, at the very least, within an important and still relevant strand of thought within biological philosophy.

Here a stark difference appears to assert itself between Nancy and Canguilhem's respective uses of the term "sense." For whereas the former uses the term to describe being in general, the latter's use appears initially only to relate to life and to the manner in which the living organism appreciates "values in relation to a need." Canguilhem seems, in his earlier 1952 work at least, to want to differentiate the necessity of sense for biological organisms from the organization of purely physical or chemical structures that are known to other sciences. Sense as an appreciation of values in relation to a need would, if extended to nonliving structures, appear to blur the specificity he confers upon life as such and to lead directly to some kind of panpsychical position according to which nonliving things would "appreciate" the value of their physical relations and "feel" needs in relation to their surrounding milieu. This does not appear to be viable from Canguilhem's perspective, and Nancy, in *The Sense of the World*, dismisses outright the possibility that his ontology of sense would open onto or imply any kind of panpsychism (1993c, 103; 1997, 62).

Yet by the time Canguilhem comes, in 1966, to align sense with the concept of information and with the existence and expression of the genetic code, things are by no means so straightforward. In this context, sense would be both the appreciation of values in relation to a need (as lived by the organism in its milieu) and the information inscribed in matter as the genetic code. The genetic code is inscribed materially and in turn is expressed, generating further material structure and engendering biological processes or activity in a complex interaction with the surrounding milieu. Given that Canguilhem in 1966 retains the notion of the living organism as a center of reference, there is therefore a strong degree of continuity between sense as lived functional relations that define its being as sense, and sense interpreted in terms of the information coded within gene sequences of the organism and the processes of gene expression. This is important because it brings together on the same ontological footing both the qualitative, lived appreciation of value in relation to need and the purely physical and biochemical structures of genes and gene expression understood as informational processes. There is no ontological separation here between the material structure of life and the physical processes that govern it on the one hand and the qualitative experience of the living organism's interaction with its material milieu on the other.

Both *are* and *are lived* as sense. It is in this context that Canguilhem understands life in general as "an activity of information and assimilation of material" (1994c, 342–43).

For both Canguilhem and Nancy, sense is matter in its processual, differential, and relational structure or organization. Once philosophical thought comes to interpret the information borne by the genetic sequences of life as sense inscribed in matter, it is not too great a step to extend the interpretation of sense so that it can refer to the structured organization of matter per se. Such a step might appear to be even more plausible in the light of Lane's argument in *The Vital Question* that there can be no clear dividing line between the material structure of a geologically active planet on the one hand and the structuring of a living cell on the other, to the extent that "geochemistry gives rise seamlessly to biochemistry" (2015, 27). In the light of such an argument, the biological organization of genetic material is simply a further extension of the tendency of matter to organize per se. The self-organization of living matter and the material structuration of nonliving matter can thus be placed on the same ontological footing and interpreted as the material inscription of sense. Self-organized living matter then gives rise to qualitative experience that makes sense for the living organism as a center of reference. In nonliving matter, it almost certainly does not. Yet the qualitative experience of lived sense arises in a seamless ontological continuity with the materiality of sense as such.[14] Both nonliving matter and material life are understood as relational structures that are, in Nancy's words, "liable to sense" (1993c, 103; 1997, 62).

On the basis of a close alignment of Canguilhem's and Nancy's thinking, the ontological continuity between qualitatively lived and materially inscribed sense allows for a fuller vision of a naturalist philosophy of sense and a naturalization of our human interaction with, and knowledge of, nature. Canguilhem speaks of his thought in "New Knowledge of Life" as a "project of naturalising the knowledge of nature" (1994b, 344; 1994d, 309). When taken together, what Canguilhem's and Nancy's thought describe is a natural material world of structures, relations, and processes where sense is inscribed in and as matter, and where living beings as centers of reference within natural milieus live or experience their worlds qualitatively as sense. It will be recalled that Nancy's ontology of "sense" describes a human-independent world of real things and objects, one that is always also a surrounding world of nature from which we as humans are never ontologically separated. This was highlighted in particular in the context of and with reference to the anticorrelationism of the speculative realists. What can now be made more explicit is the extent to which Nancy's ontology of sense can describe the being of the natural world as it would be outside of man, but can also describe at the same time the place of humans within the natural world and the continuity of the human with all other nonhuman things.

Biological thinking that has developed in the wake of Canguilhem is helpful in this regard. Jean Gayon (2004) has argued that the opposition between a metaphysical thesis that would speak about reality as it is, independent of human knowledge, and a thesis that would argue for the necessary dependency of our relation to reality on human categories is false. Although he does not put it in these terms, it is clear that from Gayon's biological perspective, the alternative of correlationism and anticorrelationism as posed by speculative realists such as Meillassoux does not in fact give an adequate account of how humans, or indeed any other living organisms, come to know their environment. In this context, he puts forward the following argument, which directly cuts across the terms of the speculative realist debate regarding correlationism: "It does not seem unreasonable *a priori* to say that, on the one hand all human knowledge is anthropocentric, that it is adjusted to human capacities and interests, and on the other hand that humans are able to construct objective knowledge, that is, knowledge the truth or falsity of which should in principle, be independent of the epistemic disposition of the objects [. . .] that refer to it" (174). In this context, Gayon argues that the debate surrounding the human dependence or independence of reality is irrelevant because "in a biological perspective, and more especially in an evolutionary perspective, what matters is not an independent world, but a surrounding world" (171). According to this view of knowledge, all living organisms, and human beings no less so, have fundamental relations with their surrounding world that constitute them as such and that ensure their development and survival. This is clearly an extension of Canguilhem's relational account of biological individuality or individuation. These elementary relations, as meaningful and constitutive, are also ways of knowing, even if they are not yet cognitive in the sense of higher processes of reasoning or reflection. For Gayon, "organisms have many windows and variables that provide them not with an image but with information that is physically and biologically pertinent for their survival" (186). Such relations need to be understood as real relations to a really and independently existing surrounding world. This is not simply an epistemological question of how a being or entity, once constituted, would gain access to an independent world. Rather, it is to say that any being or entity necessarily has access and real knowledge of a surrounding world as a condition of its survival.

If relations to the lived milieu provide false, illusory, or aberrant knowledge, an organism would not survive long. In this context, we should not place human knowledge on any kind of special footing, either by relating it to a specifically human transcendental ground or by conferring on humans some privileged access to mind-independent reality through speculative or scientific reason. Rather, "it is better to think of the relation of human knowledge to reality as a prolongation of the relation of more elementary forms of organic knowledge that is, a relation of an organism with a surrounding world" (Gayon 2004, 184). This insight is echoed within the realm of biosemiotics, a disciplinary perspective that could be developed fruitfully in relation to the problematic of "sense" being discussed here in the context of Canguilhem and Nancy. Donald Favereau, in his account of the "biosemiotic turn" within biological thinking, notes, "It is biosemiotics that will insist that in the study of biological agency of every kind, it is precisely the naturalistic establishment of *sign relations* that bridge subject-dependent experience [. . .] with the inescapable subject-independent reality of alterity that *all* organisms have to find some way to successfully perceive and act upon in order to maintain themselves in existence" (2008, 8). From the biological perspective as articulated by Gayon and the biosemiotic perspective as characterized by Favereau, the opposition between subject-dependent and subject-independent knowledge of the real that is presupposed by the correlationist thesis of Meillassoux (and by speculative realism more generally) is not only unhelpful and misleading but also entirely out of line with the reality of the way organisms, both human and nonhuman, relate to their surrounding world.

Reading Gayon's account of realism and biological knowledge in the context of both Nancy's and Canguilhem's thinking of sense (understood as "being to" and as a "relation to") allows for the referential contact of living beings with their surrounding milieu to be viewed in both ontological and epistemic terms. Not only is the being of a living organism constituted in and through its relations with the surrounding milieu but also those relations constitute the basis for real knowledge of the world. As Gayon puts it, from the biological perspective, knowledge is "a relation between real organisms in a real surrounding world" (2004, 184). Put in the terms of Nancy's ontology, it can be said that a living thing would be the sense of its relations, its being-to that constitutes it as such, but at the same time those relations would also be a form of real contact with an exteriority, which is known as such through this relational contact.

It is here that the figure of touch in Nancy's thinking of the sense of the world achieves decisive importance. For Nancy, the material differentiation and relationality of sense that constitutes the being of things is always articulated in the mode of a certain kind of touch or contact. To this extent, sense is simply touch: "Sense *is* touching. The being-*here*, side by side, of all these beings-here [. . .]. Sense, matter forming itself, form making itself firm: exaction and separation of a tact" (1993c, 104; 1997, 63). For living things, human and nonhuman alike, this mode of touch, "tact," or contact is the means by which the world comes to be known, perceived, or exposed as such. As Gayon argues, to say that what we know arises out of a relation to a surrounding world does not require us to renounce the realist conception of knowledge; rather, it leads us to "modify its spirit" (2004, 184). According to such a modified realist knowledge, specifically human cognition loses its privilege. Knowledge is a function of the elementary relations of all living organisms with their environment, and so-called higher human forms of cognition are simply an extension or prolongation of those elementary relations. To this extent, a modified realist knowledge in which human cognition loses its privileged status can never be exhaustive or totalizing because it arises from the finitude and spatiotemporal specificity of relations of sense—of contact and touch with a surrounding relational milieu.[15]

What emerges from this is a naturalized realism articulated by means of a haptic ontology of sense. In the light of this, and in the context of the alignment of Nancy's speculative thought with Canguilhem's biological thinking, the extent to which the ontology of sense is also a naturalist philosophy becomes clear. This philosophy describes a biological natural world and a relational physical universe in which the being-there of things, both living and nonliving, makes sense by itself, as such, and in excess of any correlation with specifically human beings. At the same time, the relationality of sense is the means by which living things, as centers of reference, qualitatively experience, feel, or touch on their real surrounding world and know it as such. This qualitative experience of touch or contact is both the condition for the manifestation or appearance of a surrounding world as perceived by all living beings and their source of all real knowledge of that world. Sense, Nancy writes in *The Sense of the World*, "is the movement of being-*toward*, or being as *coming* into presence or again as transitivity, as passage to presence—and therewith as passage *of* presence" (1993c, 25; 1997, 12). The touch or contact of sense, understood as an ontologically constitutive relation of existing things to other existing things, also articulates their mutual or reciprocal exposure, and with it the existence and manifestation of a world teeming with a plurality of singular entities. That such a structure of exposed existence and of world constitution can be thought entirely in terms of natural processes is consistently affirmed by Nancy in such a way that to pose any separation within his thinking in general between nature and, say, culture or technology makes no sense. That this structure of Nancean sense understood as exposed existence is entirely compatible with Canguilhem's understanding of sense in purely biological terms itself follows naturally. Canguilhem's affirmation of the very being of biological organisms as sense, and as an a priori inscribed within matter, thus combines with Nancean sense to articulate a fully materialized and naturalized understanding of the transcendental.

It has been suggested, however, that the speculative interpretation of organized matter as sense need not be restricted to the organization of living matter alone (genetic code and the information transfer it enables). If, as Lane argues, "geochemistry gives rise seamlessly to biochemistry," then there may be no reason to hold that the biological coding of life is the only way in which sense is inscribed in and as matter. Yet if this possibility is to be developed further, then a natural philosophy developed out of a Nancean ontology of sense must be brought into relation with physical or cosmological thought and not simply or solely be aligned with biological thinking. If Nancy's sense of the world describes a biological and terrestrial surrounding world, it must also describe a natural and physical surrounding cosmos.

Cosmos

Sense in Nancy, it has been argued, is never present or determinate, and the relation of sense is not a reciprocal relationship between the phenomena that form the structure of an already appearing and worldly phenomenal relationality. Rather the transimmanence of sense, anterior to phenomenal appearance, is always a relation to, where the *to* is also an exposure of a singular instance to the exteriority of another singular instance, an exposure that can never be held in reserve, ontologically disclosed, or determined as such. Yet this withdrawal and exposure of sense is always the condition of the opening of phenomenal appearance or manifestation. This set of formulations may seem eminently phenomenological, but sense here needs to be understood as real—that is, as both prior to and independent of human consciousness, as the nonsubstantive stuff of matter and things, and as the relational condition of all things appearing to each other and thus constituting the finite, exposed existence of a shared world.

In *The Sense of the World*, Nancy expresses this in explicitly cosmological terms. Yet he does so only insofar as the very idea or figure of the cosmos itself is reconfigured. The extent of Nancy's cosmological ambition is revealed in the following:

> In order to be understood as a world of sense—of "absent sense" or exscribed sense—the world must also be understood in accordance with the *cosmic* opening of space that is coming toward us: this constellation of constellations, this mass or mosaic comprising myriads of celestial bodies, their galaxies, and whirling systems, deflagrations and conflagrations that propagate themselves with the sluggishness of lightning, the almost immobile speed of movements that do not so much traverse space as open it and space it out with their motives and motions, a universe in expansion and/or implosion, a network of attractors and negative masses, a spatial texture of spaces that are fleeing [. . .], a universe of which the unity is nothing but unicity [*unicité*] open [. . .], a universe that is unique insofar as it is open on nothing, its "something" having been thrown there from nowhere, infinitely defying all themes and schemes of "creation"—all representations of production, engenderment, or mere origination [. . .], a coming always pre-vented and pre-venting, devoid of providence and yet not deprived of sense: a coming that is itself the sense (in all senses) of its starring.
>
> We do not yet have any cosmology adequate to this noncosmos. (1993c, 61–62; 1997, 37–38)

Nancy here provides a description of a cosmos whose spatial extension is opened only in and through the constellation of relations that constitute its multiple and differentiated elements. Such a universe is composed of nothing other than this opening of differentiated spatial extension.[16] It is a universe without underlying substance or ground, opening over nothing other than its own spacing, and as such, it is a singular universe. At the same time it is a universe that does not form a unified totality when taken as a whole but is rather nothing other than the fragmentary singular plurality of relational sense that constitutes it as such.

Yet Nancy is clear: the cosmos described here is not what we would normally understand in the usual use of the term; it requires an entirely new cosmology to describe it. The term "cosmos" is usually understood as the world or universe that forms an orderly or harmonious system. This is the world or universe that Newtonian science and it successors take to be governed by timeless natural laws—laws discovered by modern science, by its experimental methods, its theories and mathematical descriptions, and laws that, insofar as they do describe a universal order, would apply at all times and at all places in the universe. To this extent, the perspective of modern science that would discover universal and timeless cosmological laws is one that places itself outside of the universe as a whole and looks into or over it as a total object under an all-seeing gaze. Nancy suggests that this is an essentially theological, or ontotheological, point of view (the collective knowledge of science and laws of nature being akin to the knowledge of God), and that the universe or cosmos viewed as the spacing or opening of sense cannot only be described from the perspective of a panoptical, scientific subject that has the totality of a harmonious and orderly universe as its object. In a formulation that both echoes and roundly rejects Descartes's thinking of the scientific subject in the *Discourse on Method*, Nancy writes: "We would have to begin by disengaging ourselves from the remains of the old cosmo-theo-ontology, such as these remains still supported a 'conquest of space' conceived [. . .] in terms of a *kosmopoiesis*: mastery and possession of the universe" (1993c, 62; 1997, 38). Nancy here suggests that a cosmos, opened in its spatial extension and in the relationality of absent, withdrawn, or exscribed sense, is not fully compatible with the rationalist metaphysics that underpinned the birth of modern scientific humanism in the age of Newton and Descartes. The horizons of unity, totality, order, harmony, and mastery that governed such metaphysics must be abandoned in any cosmology that would be adequate to the acosmos of relational sense.

The Withdrawal of the One

In *What's These Worlds Coming To?*, coauthored with French cosmologist and astroparticle physicist Aurélien Barrau, Nancy argues that contemporary science is indeed inventing a new cosmology, one that is in the process of divesting itself of inherited metaphysical prejudices: "It so happens that the *cosmos* that is studied or invented by cosmophysics today no longer corresponds to this depiction of an absolute and unitary harmony" (2011, 13; 2014, 2). In such a physics, the horizon of unity or the figure of the One is to be abandoned, and with this, "the myth of a real that is a unitary-unity finds itself undermined" (2011, 19; 2014, 6). These statements might seem strange, counterintuitive, or self-contradictory; surely cosmology is necessarily the study of the universe taken as a unity or whole. It is the study of the origin, evolution, and general structure of the universe and the universal laws that govern it. Yet this, precisely, Nancy argues, is the meaning of cosmology and the image of the cosmos that is in the process of being transformed.

From the perspective of his singular plural ontology of sense, the whole of the universe, if it is to be understood according to the figure of the One, cannot be grasped as a unity from the point of view of a subject who would stand outside of, or above, the universe in a position of mastery. Rather, the One of the universe would need to be understood as prior to and radically withdrawn from any graspable horizon of totality or unity. As Nancy himself puts it: "It might be in effect that the One prior to any unity escapes not only our grasp—which it has always done as we have always known—but withdraws itself from itself" (Nancy and Barrau 2011, 35; 2014, 17).[17] If the whole of the universe were to be understood as One, then this must be a One that is always in excess of itself, a One without self-identity, whose multiplicity and self-differentiation offer no position, perspective, or ground from which the whole could be grasped or articulate itself as such. This would be a paradoxical One, withdrawn from all figures of unity that structure given identities in an already appearing or manifest world. It would be a One radically withdrawn from the multiplicity of phenomenal appearance, yet a One that is at the same time the condition of the opening of any and all appearance: a "One from beyond or from before the One. It is the punctual One without dimension. It is that which has not taken place—neither place nor time. Rather, it opens the possibility for time and place" (2011, 36; 2014, 17). In a move that recalls and repeats the key difference between Nancy and Laruelle identified in chapter 1, it becomes clear once again that Nancy is thinking the immanent or transimmanent structure of the real as material differentiation that opens, spaces, and disperses all the things or elements that make up the cosmos in the radical absence of any overarching unity or order.

Nancy's physicist coauthor, Aurélien Barrau, agrees with this view and argues that contemporary science is far from presenting a unified image of an orderly and harmonious cosmos. Yet he also highlights the extent to which this runs against the grain of our traditional understanding of physics as a discipline—and indeed against the clear evidence of the unity and coherence that it has established in its evolution as a discipline over time. Physics, he notes, has created an extraordinarily coherent edifice that has accomplished a conceptual and normative unification that has succeeded in subsuming "the chimeric diversity of the real under a reduced number—which is inescapably aimed toward unity—of principles" (Nancy and Barrau 2011, 63; 2014, 34). To this extent, physics appears to be, both in its essence and in its collective and cumulative practice, "unitary in its essence" (2011, 63; 2014, 34). Yet he argues that contemporary physics observes multiple and radically divergent phenomena; it is shot through with theories, which all appear to defy subsumption into any set of unified or overarching physical laws. Here he cites the phenomenon of symmetry breaking, the multiplication of free parameters within supersymmetry, the generation of a quasi-infinity of effective laws within string theory, and theoretical hypotheses regarding the existence of a multiverse. In the light of all such instances, he argues, physical laws can perhaps only be understood as "simple environmental parameters" and therefore as a "contingency lying at the heart of formal necessity" (2011, 65; 2014, 35). Inscribing local contingency within the idea of physical law immediately undermines its status as a universal regularity governing the cosmos as a whole and thus guaranteeing the unity and orderliness of that whole. This leads Barrau to conclude that although the development of physics may well have been motivated and propelled by the desire for the unification of natural phenomena, this does not at all mean that the "physical-mathematical approach inevitably leads to a unitary real, a world unity, or a uni-verse" (2011, 66; 2014, 35).

This is confirmed, he argues, not just by the existence within contemporary physics of a multiplicity of incompatible or heterogeneous theories that have their own domains and that describe different levels of reality (most notably general relativity and QM), but also by the existence of different and incompatible theories that seek to explain the very same phenomena. Here he cites the divergent approaches to the problem of quantum gravity: "Strings, loops, topoï, non-commutative geometry, causal triangulation, path integrals, twistors, quantum geometrodynamics, and entropic force are many possible mediations of quantum imperatives with gravitational invariants. All of these approaches essentially agree with scientific observations" (Nancy and Barrau 2011, 67; 2014, 36). One or another of these diverse and divergent theoretical approaches to quantum gravity may well be found to be the best theory in the light of future experimental observation. Yet Barrau suggests that if there is one thing that epistemologists agree on, it is "theories' mortality, temporality, and ephemerality—not to mention their precariousness" (2011, 68; 2014, 37).[18] One of the currently competing theories of quantum gravity may win out in the near or medium term, but in the long term, it will no doubt be superseded by further theory. This leads Barrau to conclude that "propositions, and this is a defining property of science, will necessarily be contradicted in the more or less distant future" (2011, 68; 2014, 37). It seems that the number of competing theories shows no sign of diminishing with time. As some are eliminated through experimental results, others are born, and this, Barrau suggests, leads science to point the way toward an irreducibly plural real (2011, 112; 2014, 63). Thus, in the light of both the diverse multiplicity of physical phenomena and the ongoing proliferation of divergent theories of physics, the figure of a unified science and a unified and orderly cosmos cannot be sustained.

Aurélien Barrau, of course, is just one scientist, albeit one who has received significant recognition at quite a young age for his work on black holes, quantum gravity, and primordial cosmology, and one who has also written a number of philosophically oriented books on fundamental questions in physics.[19] However, the arguments he puts forth in *What's These Worlds Coming To?* cannot necessarily be said to reflect a consensus among contemporary scientists and philosophers of science. They can nevertheless be broadly aligned with distinct and established positions within philosophy of science, such as those of the Stanford School theorists, who argued for the disunity of science in the context of an ontological pluralism and realism.[20] More significantly from the perspective of this argument, both Barrau's and Nancy's image of the cosmos and the invocation of a new cosmology are mirrored in key respects in the writing of American theoretical physicist Lee Smolin and his long-term collaborator, Brazilian philosopher Roberto Mangabeira Unger. In works such as *The Life of the Cosmos* (Smolin 1997), *Time Reborn* (Smolin 2013), and *The Singular Universe and the Reality of Time* (Unger and Smolin 2015), an extended critique of contemporary physical and cosmological theories is developed and a polemic is launched against the overvaluation and privileging of mathematics within modern physics more generally.

Temporal Naturalism and the Godlike View

Smolin, like Nancy (and indeed like Ladyman and Ross), argues for a fully relationist picture of the universe. Like Ladyman and Ross, he cites the authority of modern and contemporary scientific theories in order to affirm this picture, according to which structure and relation, as well as the organization of matter out of structure and relation, account for all that exists as such. In *The Life of the Cosmos*, he points out that all physical theories today are "based on the point of view that the properties of things arise from relationships" (Smolin 1997, 51). In the debate I staged earlier between Nancy and Harman concerning things, relations, and relata, Smolin is unequivocally on the side of Nancy and against Harman: "All properties of objects are based on relationships between real things and have no absolute meaning" (54). On this point, he directly cites the phenomenon of quantum entanglement, arguing that the "entangled nature of the quantum state reflects something essential in the world" (252). Quantum entanglement tells us that the properties of matter at a fundamental level are deeply interdependent and that the idea of a principle of separation or isolation at work within the deep structure of reality (e.g., atomism or, indeed, Harman's vacuum-enveloped objects) runs against what has been discovered experimentally about that structure. For Smolin, the empirical verification of quantum entanglement is "one of those rare cases in which an experiment can be interpreted as a test of a philosophical principle" (252).

Smolin continues this defense of relationism in his more recent work, arguing in *Time Reborn* that "every entity in the universe evolves dynamically in interaction with everything else" (2013, 116). In *The Singular Universe*, he modifies this position slightly, conceding that temporal physical events have some intrinsic properties as well as relational properties (Unger and Smolin 2015, 483).[21] However, what is fundamental to the positions adopted in *The Life of the Cosmos*, *Time Reborn*, and *The Singular Universe* taken together is the argument that the universe as a whole needs to be understood temporally in its historical evolution and relational becoming, and according to the paradigms that inform the biological understanding of phenomena. What matters from Smolin's relationist cosmological perspective in each of these works is "whether we can *understand* the whole of the universe as comprising a single, interrelated system" (Smolin 1997, 13). If the whole of the universe can be understood as a singular system or structure evolving in time, Smolin argues, then the "laws of nature themselves, like the biological species, may not be eternal categories, but rather the creations of natural processes occurring in time" (18). Earlier it was shown that a contemporary biologist such as

Nick Lane can argue that biological life must not be separated or divided in any significant way from geophysical and geochemical processes (2015, 27).[22] Here we see a contemporary theoretical physicist arguing that the evolution and development of the physical universe, together with the lawlike regularities produced within it, must be understood as something akin to the evolutionary processes of biological life. The consequence of this insight into the temporal evolution of the cosmos understood as an interrelated structure or system is a radical calling into question of the preeminent role played by mathematics within modern physical and cosmological theory.

Like Nancy, Smolin argues that scientific theory and practice have been underpinned by perspectives that are essentially theological in nature. He explicitly identifies this tendency with the privileging of mathematics and traces a trajectory that runs from antiquity to the present day. "From Pythagoras to string theory," he suggests, "the desire to comprehend nature has been framed by the Platonic ideal that the world is a reflection of some perfect mathematical form" (1997, 177). Already in *The Life of the Cosmos* Smolin's theory of cosmic evolution challenges the idea that the deepest structures of nature have anything to do with mathematics (1997, 179). Yet it is in *Time Reborn* that he launches his full polemic against the privileging of mathematics within physics. In his preface, he writes: "If we believe that the task of physics is the discovery of a timeless mathematical equation that captures every aspect of the universe, then we believe that the truth of the universe lies outside the universe" (2013, xvi). This, precisely, is the theological perspective identified by Nancy in *What's These Worlds Coming To?* and described as the panoptical position of a scientific subject whose knowledge of the cosmos, deriving from a point of view outside of the universe and grasping it as a whole, would be akin to that of God. Yet if, as Smolin argues, the universe is nothing other than a singular relational structure evolving dynamically in time, then there can be no such timeless perspective on the cosmological whole from the outside. The, in essence, Platonic and theological view of mathematics as a reflection of the timeless extracosmic truth of the universe was carried over into modernity, Smolin argues, in Newton's theories and in what he calls the Newtonian paradigm.[23]

The Newtonian paradigm, Smolin argues in *Time Reborn* (2013), is the method by which scientific experiment isolates subsystems of the universe and concentrates its attention on a limited number of variables, objects, or particles within that subsystem. He calls this "doing physics in a box," and the decisive moment in such an approach is "the selection, from the entire universe, of a subsystem to study" (39). This subsystem may be the space in which an object falls according to gravitational forces or any other number of systems (planetary systems, quantum systems, and so on) that can be isolated as such, and in which variables can be determined and measured. The point here, Smolin suggests, is that such an artificially isolated system will always be "an approximation to a richer reality" (39). The subsystem comprises all the variables that must be measured for the system itself to be known at any given point in time. This list of variables constitutes the configuration of the system, and the set of all possible configurations is represented in an abstract space known as the configuration space. The very act of isolating and determining a system according to its configuration and representing it according to a configuration space is inescapably an abstraction and a selective procedure such that what is measured and represented will always be "an approximation to a deeper and more complete description" (39).[24]

Much is thus necessarily left out when physics is carried out according to the procedures of abstraction, isolation, and measurement that constitute the Newtonian paradigm. Crucially, what is also left out, Smolin argues, is the clock used to determine the point in time in which measurements of the subsystem is made. As Smolin puts it: "The clock is not considered part of the subsystem, because it's assumed to tick uniformly despite whatever is going on in the subsystem" (2013, 40). To this extent, the configuration space itself is without time. Each point of time would be measured externally and represented as a curve within the configuration space itself, and thus as a timeless spatial or geometrical object. The perspective of the Newtonian paradigm, Smolin argues, is thus once again that of a God standing outside the system under observation and making measurements according to an absolute notion of time that is also external and therefore not part of the subsystem of the universe being measured. This, he points out, is entirely consistent with the theological commitments of Newton himself, who believed in "the fantasy that the physics he invented captured God's view of the universe as a whole." It took the Einsteinian revolution to find a way of putting the clock back into the fabric of the universe itself with the notion of a space-time continuum (40–41). Yet time, in Einstein's theory (and in particular the block universe picture and Minkowski space-time), is still spatialized insofar as it is considered to be a further dimension of space itself. Time as a dimension of space in the block universe image is represented as yet another timeless geometrical object. At their most basic level, then, special and general relativity are also theories of timelessness (55).

What the abstracted configuration space of the Newtonian paradigm and the Einsteinian conception of space-time share, Smolin argues, is that they both represent the world and its temporal history by means of a mathematical object that spatializes time and renders the world in effect timeless. "The space-time of general relativity corresponds to a mathematical object much more complex than the three-dimensional Euclidean space of Newton's theory," he notes, but it is still "timeless and pristine" (2013, 71). As its title suggests, the argument of *Time Reborn* is that the fundamental reality of time and temporal becoming needs to be reasserted within the paradigms of physics and cosmology. At the heart of this evacuation of time from physical theory is the metaphysical privileging of mathematical object or models in the description of the physical universe. As Roberto Mangabeira Unger has argued with Smolin in *The Singular Universe*, what mathematics gives us is a representation of the physical world "eviscerated of time and phenomenal particularity. [Mathematics] is a visionary exploration of a simulacrum of the world, from which both time and phenomenal distinction have been sucked out" (2015, 15). Smolin's relational cosmos is one that is always evolving historically, fundamentally, and singularly in time. The Newtonian paradigm, with its subsystems and timeless configuration spaces, thus cannot be a means of representing the cosmos as a whole. Neither can the complex mathematics of Einsteinian relativity theory. As Smolin puts it: "Logic and mathematics capture aspects of nature, but never the whole of nature. There are aspects of the real universe that will never be representable in mathematics. One of them is that in the real world it is always some particular moment" (2013, 246).[25] It is because mathematics always eviscerates the reality of spatial and temporal particularity, and because relational reality is always and necessarily temporal at its most fundamental level, that the Platonic image of the universe as corresponding to a timeless mathematical object can never be correct.

In both his single-authored works and in his collaboration with Unger, Smolin consistently argues that physics and cosmology must embrace time as a fundamental aspect of reality, and that in order to do so science must reevaluate the status it confers on the mathematical description of the universe. Mathematics can remain an absolutely necessary and extremely powerful descriptive tool, a tool deployed within theoretical constructs and models that should be both predictive and verifiable or open to both falsifiability and confirmation. Yet mathematics should never be held up as a royal road or exclusive gateway to a timeless truth that would somehow be situated outside of the universe as its ultimate and eternal reality. In other words, mathematics is stripped of the metaphysical and ontological privilege conferred upon it in the Pythagorean vision in such a way that "a deep chasm opens up between nature and mathematics" (Unger and Smolin 2015, 141). This position leads Unger and Smolin to develop the beginnings of a philosophy of nature that they dub "temporal naturalism." The role of temporal naturalism would be to give a philosophical description of the universe that is stripped of the mathematicorationalist and quasi-theological metaphysics that, they argue, has informed physics since at least the time of Newton. In so doing, it would allow both physics and cosmology to see beyond the inherited prejudices that have led science into a crisis of the naturalist worldview and that have inhibited the forward progress of physical science as such. A new temporal naturalism would thus open up the possibility of novel futures for both theoretical reflection and for experimental practice. Unger and Smolin put this as follows: "What natural philosophy can and should do, in its role as the scout of science and enemy of the metaphysical and methodological preconceptions that restrain its progress, is to foreshadow theory" (2015, 87). Temporal naturalism would therefore demand a critical stance vis-à-vis existing theories and a separation of experimental findings of science from the "metaphysical pre-commitments in the light of which the significance of these findings is commonly interpreted" (163).

It is worth noting at this juncture that both Unger and Smolin are widely considered to be mavericks within their respective fields of cosmology and philosophy. Smolin in particular is a highly controversial figure within the field of contemporary physics. His book *The Trouble with Physics* (2007) caused a great deal of polemic as a result of its critical evaluation of string theory, and it was criticized for misrepresenting the theory itself and for presenting a number of misconceptions in relation to it (Carroll 2006; Hossenfelder 2007; Polchinski 2007). Given this context, it is worth assessing some of the key positions of Unger and Smolin's temporal naturalism in the context of wider debates within the philosophy of science and mathematics. The two key issues here relate first to the fundamental status Unger and Smolin accord to time and to the notion that the laws of physics may be said to evolve, and second to their anti-Pythagoreanism and question of the status of mathematical forms and their relation to physical reality.

Smolin's rejection in *Time Reborn* (2013) of the tendency of modern physics to render the world timeless and to configure physical laws in terms of a "higher" eternal reality is not in any way a novel or unique position. Physical chemist and Nobel laureate Ilya Prigogine developed similar views derived from his work on thermodynamics, dissipative structures, and complex systems. These are expressed clearly in *Entre le temps et l'éternité* [Between time and eternity], published in collaboration with Isabelle Stengers in 1988. Like Smolin, Prigogine and Stengers affirm that both QM and general relativity inherit directly from classical dynamics and give a timeless image of the cosmos insofar as they negate the irreversibility of time by describing deterministic processes which are in principle reversible (9, 126, 164). In this context, they also, like Smolin, discern a fundamental Platonism to be the underlying metaphysical prejudice of much modern physics (explicitly so in Einstein's writings) insofar as an opposition is affirmed between atemporal or eternal fundamental laws and the mutable realm of phenomena (228). Against this Platonic attitude, Prigogine and Stengers argue for the existence of a fundamental "arrow of time" that underpins all phenomena and the deep structure of the universe as a whole. Prigogine and Stengers's universe is a nondeterministic, probabilistic universe in which "irreversibility is inscribed in nature" and the "laws" of nature "concern probabilities of evolution in a future that they do not determine" (116, 28). (Here they perhaps echo Barrau's argument alluded to earlier that physical laws need to be understood as "simple environmental parameters.") The striking proximity of Smolin's arguments to those of Prigogine and Stengers suggests that they are not simply the result of the fanciful, speculative musings of a lone maverick or dissident but rather have a legitimate place within wider scientific debate.

Indeed, as American physicist Abner Shimony has pointed out, Smolin is in good scientific and philosophical company in arguing for a temporal evolution of the fundamental laws of nature, with proponents of this view including the likes of Charles Sanders Peirce, Alfred North Whitehead, and John Archibald Wheeler. Shimony is sympathetic to the view that fundamental physical laws might evolve in time but is ultimately skeptical, arguing that the processes of such evolution must ultimately and necessarily evolve against a more fundamental background that nevertheless does have basic properties that themselves do not evolve (Butterfield and

Pagonis 1999, 220).[26] Shimony's qualified combination of sympathy and skepticism is echoed in a different way in the thought of British philosopher John Randolphe Lucas, who published extensively on questions concerning space, time, and space-time. Lucas is broadly in agreement with Unger and Smolin's rejection of the idea of a timeless perspective on the cosmos. He argues that the Platonic "view from 'Nowhen'" according to which both scientist and scientific knowledge are abstracted from spatiotemporal particularity is "powerful but misguided" (Butterfield 1999, 3). However, although Lucas does concede that much twentieth-century physics appears to have required a "tenseless account of time" (e.g., in the block universe image and in Minkowskian space-time; Butterfield 1999, 6), he nevertheless argues that ultimately modern science offers a picture according to which the anisotropy (i.e., irreversibly directional) nature of time is deeply embedded in the structure of the universe and that this is particularly evident in QM (Butterfield 1999, 11). Here he disagrees with Unger and Smolin (2015) and with Prigogine and Stengers (1988), arguing that QM, however deterministic its equations, offers a probabilistic vision where the before and after of a measurement articulate a fundamental directional flow of time (this being a specific, and he concedes potentially controversial, interpretation of the quantum measurement problem) (Butterfield 1999, 14).[27] Although Lucas, like Smolin, Unger, and Prigogine and Stengers, agrees that the directional flow of time needs to be viewed as fundamental within the physical universe (and thus implicitly also rejects the Platonic or Pythagorean vision), he also disagrees with them insofar as he argues that this view of time is more widely embedded in modern physics than they would all concede.

What this brief overview of similar but divergent positions suggests is that despite Smolin's controversial status as a troublemaker within the physics community, his and Unger's theses regarding the reality of time and the evolution of physical laws are by no means unique or excessively heterodox. They do, as suggested earlier, enjoy a legitimate position within the wider field of philosophical or scientific argument and debate. Perhaps more controversial would be Unger and Smolin's polemic against the metaphysical status accorded to mathematics in physical science. Yet as has been shown, their critique of mathematics is intimately bound up with the concomitant affirmation of a fundamental reality of time and of a singular universe. From the perspective of the philosophy of mathematics, however, one wonders how controversial their anti-Platonism or anti-Pythagoreanism really is.

Unger and Smolin's (2015) account of the role played by mathematics in modern physics places them squarely within debates central to the philosophy of mathematics regarding independent existence or otherwise of mathematical objects and the relation and applicability of these to the physical universe. Not surprisingly, their account pits itself uncompromisingly against mathematical Platonism, understood in its broadest and most classical sense as the belief that mathematical forms enjoy an abstract and objective existence independent of human beings, who, in the work of mathematics itself, somehow discover them in a realm that exists apart from or beyond spatiotemporal becoming and particularity.[28] They offer extended engagements with mathematics and with the question of the role it plays in describing physical reality. They argue that mathematical objects or forms are not independently existing but are created—or as they put it, "evoked"—in the activity of mathematics itself. As with the rules of chess and the potentially infinite possibilities they create, once the axiomatic system of a particular form of mathematics has been evoked, it can be explored, developed, extrapolated, and even "discovered" as if it has an independent life or status of its own (Unger and Smolin 2015, 425). Being a result of creation or evocation means, however, that mathematics does not have the privilege of describing or providing access to a higher, atemporal realm of Truth. Rather, it is an extension, albeit one that is highly disciplined and formally rigorous, of a wider human rational activity that seeks to understand the world. This view appears to place Unger and Smolin broadly but firmly within the realm of intuitionist and constructivist accounts that hold that human-authored proofs and constructions are all there are in mathematics. It also echoes the fundamental idea of Imre Lakatos that, as Ian Hacking puts it, "mathematicians construct what they discover. Their work becomes 'objective' as it becomes alienated from their productive activity" (2014, 31). Lakatos himself articulates this in his famous and influential work *Proofs and Refutations* in the following terms: "Mathematics, this product of human activity, 'alienates itself' from the human activity which has been producing it. It becomes a living growing organism" (1976, 146). Once again, it appears that Unger and Smolin's position, although it may well be vehemently opposed by Platonists committed to the objective existence of abstract mathematical entities, is not in itself particularly wayward in the context of wider philosophical debate. The broad notion that mathematical reality is constructed or evoked by humans is perfectly respectable, forming the specific tradition of intuitionism and constructivism within the philosophy of mathematics and held in specific ways by the likes of Lakatos, also held by figures such as Richard Dedekind before him, and looked on favorably by philosophers of science such as Ian Hacking after him (2014, 236).

The same is arguably true for Unger and Smolin's ideas about the applicability of mathematics to the universe or cosmos in physics. The central point of debate here turns around the effectiveness—what Eugene Wigner (1960) famously called the "unreasonable effectiveness"—of mathematics in successfully describing physical reality. This effectiveness has often been taken as key argument for the objective and real existence of mathematical entities independent of humans (and therefore for a Platonist understanding of mathematics as well as a strongly realist understanding of science). Wigner notes that "the enormous usefulness of mathematics in the natural sciences is something bordering on the mysterious and that there is no rational explanation for it" (1960, 2). He goes on to describe this usefulness as nothing short of a "miracle" (7) and suggests that "the laws of nature must already be formulated in the language of mathematics to be an object for the use of applied mathematics" (6). This might immediately suggest a Pythagorean view of the universe of the kind that Unger and Smolin would oppose. Against Wigner, Unger and Smolin argue in *The Singular Universe* that the effectiveness of mathematics in physics is in fact quite reasonable to the extent that "wherever it is effective, there is a reason for it" (2015, 428). This implies an understanding of mathematics as offering a model or mapping of reality, where the mapping of abstract mathematical structures onto target physical structures implies a mirroring of selected structural features but not a perfect isomorphism between the mathematical and the physical, nor a totalizing description of one by the other. Unger and Smolin put this as follows: "It will always be the case that the use of mathematics to model nature will be partial—because no mathematical object is a perfect match for nature. [. . .] Those mathematical objects that provide partial mirrors of the world are a small finite subset of the potentially infinite number of mathematical objects that might be evoked. So the effectiveness of mathematics is limited to what is reasonable" (428). One could argue that this mirroring of model and target structures may remain reasonable (rather than miraculous), even when predictions made by the theoretical mathematical model are subsequently confirmed by empirical verification (as has consistently been the case with the most successful theories, such as relativity and QM). This need not testify to the deep mathematical reality of the universe but simply to the interrelatedness of structure as structure. The fact that one part of a mathematical model successfully describes one part of the empirical-structural field means that it should not be surprising if other structurally related parts of that model subsequently describe other related parts of the empirical-structural field—ones that were not discovered as testable at the time the mathematical model and its accompanying theory were initially created or evoked.

This mapping account of the applicability of mathematics within physics is, once again, a perfectly legitimate position within the philosophy of science, even if, once again, it might be hotly disputed by physicists with Platonist or Pythagorean leanings. What is perhaps controversial about it is the fact that Smolin (2013, 39) insists that mathematical objects can only ever provide partial and never total models for physical reality and that something may necessarily always be omitted by such models. This follows on naturally from the anti-Pythagorean conclusion that the singular temporal existence of the universe is not mathematical or eternally fixed in its deep structure but always evolving in a multiplicity of dynamic becoming, a becoming that will always be in excess of the formal descriptive frames that are brought to bear upon it. This is echoed in Unger and Smolin's joint claim, cited above, that mathematics offers descriptions "eviscerated of time and phenomenal particularity" (2015, 15) and Smolin's claim in *Time Reborn* that "logic and mathematics capture aspects of nature, but never the whole of nature" (2013, 246). One might object that mathematics can and regularly does describe or model dynamic temporal processes (e.g., in fluid dynamics or more generally in the mathematics of dynamic systems theory), but this, Smolin would no doubt respond, always occurs in the context of "doing physics in a box" (i.e., in the context of isolating subsystems). Such an objection therefore does not detract from the fact that applied mathematical description is always necessarily an abstraction from the particularity of the reality it describes. Indeed, this is arguably the very source of its applicability as such to the physical world. As Ian Hacking remarks, when it comes to the "the core of the applicability of mathematics," we "have learned that when we do mathematics, being with a capital 'B' is disregarded. [. . .] This abstraction, a radical abstraction, is what gives mathematics both its theoretical power and its rich connections to reality" (2014, 257).

It should be clear from all this that, without arguing that Unger and Smolin's polemical positions are somehow definitively correct, it is possible to conclude that they are perfectly legitimate within the wider context of philosophy of science and mathematics. The broad philosophical scope of their arguments and their criticisms of what are perceived to be widespread and entrenched metaphysical prejudices of modern scientific theory are bound to be controversial and subject to further debate among both scientists and philosophers. Yet they are arguably following in a tradition of critique of scientific thought that is embodied in the likes of Whitehead, who argued forcefully in *The Concept of Nature* that the scientific expression of certain basic facts about reality "has become entangled in a maze of doubtful metaphysics" ([1920] 2015, 17). One might argue more generally that, just as there has been a tradition within Continental philosophy after Nietzsche and Heidegger of overcoming Platonic idealism and destroying or deconstructing metaphysics and notions of eternally static and grounded Being, so Unger and Smolin are continuing a comparable tradition within the realm of philosophy of science, one that is embodied in the work of figures such as Whitehead and Paul Feyerabend. The aim is above all to strip science of its traditional metaphysics (the realism of matter, substance, and subject–object or subject–predicate relations) in order to develop different ways of conceiving or imagining the temporal reality of a singular and always dynamically evolving universe.

It may not be surprising, therefore, that the picture of the cosmos that Unger and Smolin's temporal naturalism affirms is, in key respects, close to that put forward by Nancy in *The Sense of the World* (1993c, 1997) and by Nancy and Barrau in *What's These Worlds Coming To?* (2011, 2014). In many ways, their vision closely resembles a cosmology that would be commensurate with the acosmos that was described by Nancy in *The Sense of the World*. Nancy, remember, called for "an a-cosmic cosmology that would no longer be caught by the look of a *kosmotheoros*, of that panoptic subject of the knowledge of the world" (1993c, 62; 1997, 38). The singular universe described in Unger and Smolin's temporal naturalism is one in which the external timeless perspective afforded by the privileging of mathematics is removed. The structures, organized matter, and lawlike regularities of the physical universe are not grounded in a timeless external reality described by mathematical objects but rather result from nothing other than the historical becoming of structure itself (Unger and Smolin 2015, 13) such that there simply is nothing other than dynamic structure. Any attempt to think, know, or grasp the reality of structure must come from a position that is internal or immanent to the temporal becoming of that structure itself.

This necessarily and once again dissolves all horizons of totality and unity that would govern cosmological science and radically calls into question all aspirations to ultimate mastery of nature. It indicates that there must be limits to the powers and possibilities of science and theoretical knowledge per se in relation to the whole of the cosmos or the totality of the real. The impossibilities or antinomies generated by the mathematically oriented image of an eternal universe when it is confronted with the irreducible and fundamental temporality of the actually existing universe may be "a sign not only of the limits to the powers of science and of its ally in natural philosophy but also of our groundlessness—our inability to grasp the ground of being or existence" (Unger and Smolin 2015, 102). Nancy and Barrau similarly argue that the abandonment of any idea of an eternal world beyond this world means that the limits of the universe may not be graspable as such: "These limits themselves are only given with the caveat that it is impossible to properly assign them as delimitations of a world in relation to what is beyond or behind it" (2011, 91; 2014, 50). Without any purchase on an outside and against which the limits of the universe can be traced, those limits cannot be themselves grasped as such. Yet this experience of ungraspable limits should not be confused or conflated with the delimitation of human knowledge in relation to a hidden noumenal realm of things "in-themselves." As was the case then, and as Unger and Smolin's comment cited above confirms, the experience of limits here is an experience of the groundlessness of thought and existence and of a subsequent impossibility of grasping or determining the outer limit of all that exists in such a way that would allow it to be mastered as a totality or whole from a perspective that would be external to that existence. The limitation here relates to the possibility of locating an ontological or metaphysical ground of the universe (as substance or essence) and of grasping a cosmological unity or totality from a God's-eye view. As was argued in relation to the discussion of Gayon and Nancy in the preceding section, the knowledge of entities and things by other entities and things (including but not restricted to human entities) should be understood as a function of real relations to a real, existing surrounding world with which entities and things are always in a relation of contact and separation. In this context, the notion of an explicitly human limitation of intersubjective knowledge in relation to a noumenal realm no longer makes sense. Yet with the loss of an eternal world encasing, enveloping, or subtending, the contingent becoming of this surrounding, real, and relational world means that its temporal becoming is without ground and without determinable limit, and therefore can never be exhaustively known as a total object of knowledge.

Despite their shared view of a cosmos stripped of any external godlike perspective developing dynamically as pure relational structure, there may appear to be a key difference between Nancy's acosmology of sense and Unger and Smolin's temporal naturalism. This difference relates to the question and fundamental status of time itself. Whereas the whole point of Unger and Smolin's thinking is to argue for time as a unidirectionality of becoming that is fundamental to the structure of the universe, Nancy appears to reintroduce the language of eternity when talking about the acosmos of sense. Nancy is always interested in the opening of a world or worlds in terms of the "coming to presence," a world or worlds that ceaselessly come to presence in the circulation of or being-to of sense. So, for instance, in *The Sense of the World*, he writes: "There is only eternity as the spacing out of every present of time" (1993c, 108; 1997, 66). Or, as was shown earlier in *What's These Worlds Coming To?*, Nancy describes the nontotalizable and nonunified "One" of the universe as "the punctual One without dimension. It is that which has not taken place—neither place nor time. Rather, it opens the possibility for time and place" (Nancy and Barrau 2011, 36; 2014, 17). He goes on to add decisively that the reticulated structuration of cosmological sense (what he terms "struction" in French) "opens onto a temporality that definitely cannot correspond to a linear diachrony" (2011, 95; 2014, 52). Such an emphasis on an "eternity" of the spacing of sense that opens presence, on a timeless "punctual" universe, and on a resolutely nonlinear temporality could surely not be more different from Unger and Smolin's insistence on the fundamental status of time.

Yet the difference or opposition here is not as straightforward as might initially seem to be the case. This becomes more apparent in the light of a key difference between Unger and Smolin. Although their project of temporal naturalism is jointly elaborated and coauthored, they nevertheless acknowledge and highlight differences in their respective positions. They underline the importance of these differences and devote a whole section at the end of *The Singular Universe* to enumerating and evaluating them (Unger and Smolin 2015, 512–32), but at the same time they note that they are not decisive in relation to the overall coherence and shared aims of their project. One such difference relates to their respective view of the structure of the cosmological temporality that each holds to be so fundamental. Smolin argues that scientific thought and practice requires an objective distinction between past (as having been real), present (as real), and future (as a real to be) (522). Unger, on the other hand, rejects this schema, and in particular the doctrine of presentism (the idea that we live through a succession of "nows" which become past "nows" and in the expectation of future "nows"). In place of both presentism and its alternative "eternalism," Unger and Smolin favor a model that emphasizes fundamental temporality as a mode of becoming (518–21). They reject the category of the "present" insofar as they hold that any given "now" is never entirely self-present but rather is always an instance of becoming that overflows the identity and unity of a "now": "Presentism is untenable. Reality fails to remain within the now [. . .]. Everything in the universe is always becoming, or ceasing to be, and changing into something else [. . .]. This process of becoming and of ceasing to be is not only real, it is more real than anything else in nature. It is real if anything is [. . .]. It cannot, however, be accommodated within the now because the now is instantaneous" (247). To this extent, Unger's understanding of time can be broadly aligned with other philosophical accounts of temporality understood as a becoming or duration in which the present is stretched between past and future, between retention of what has been and the anticipation of what will be (such as may be found in Nietzsche, Bergson, and phenomenological accounts of temporality). The present therefore lacks the self-identity or self-sameness that might be attributed to any metaphysics of presence or to a philosophical or scientific "presentism."

Unger's position echoes Nancy's rejection of a linear and diachronic temporality that would be based on a succession of present "nows." Indeed, arguably the language of "eternity" used by Nancy needs to be understood much in terms of a model of "becoming" rather than in terms of any kind of static or unchanging temporality. So, for instance, in *Being Singular Plural,* Nancy once again describes the circulation of sense, as it constitutes the reticulated and relational structure of singular plural being, as "eternity" (1996, 21–22; 2000, 4). Yet he does so here, as he does elsewhere, with reference to the Nietzschean doctrine of eternal recurrence. This sets his thinking squarely within a post-Nietzschean understanding of existence as a flux or flow of becoming rather than as a static substance or essence. So Nancy's use of the language of eternity in fact articulates a rejection of a linear, chronological succession of present and self-present instances in favor of a flux of becoming that would be prior to, and at the same time make possible, the experience of temporality as successive duration. To this extent, the being-to of relational sense in Nancy always needs to be understood both as a spatialization and as a temporalization—that is to say, the dynamic becoming of the reticulated structure of sense constituted in and by the prepositional "to" of existence.[29]

Arguably, then, the difference here is between Nancy and Unger's understanding of time as becoming and Smolin's belief in the necessity of an objective reality of past, present, and future. As indicated above, both Unger and Smolin accept their differences as legitimate divergences with their shared project of a temporal naturalism. Whether temporality be viewed in terms of a flux of becoming (Unger/Nancy) or as the reality of a flow of time, what is most important here is the image of the universe, found in both Nancy's acosmology of sense and in Unger and Smolin's temporal naturalism: a singular, relational universe, whose structures and networks of relations evolve in the flux of time and in the absence of timeless laws, godlike perspectives from the outside, or any possibility of a unifying foundation or ground.

Speculative Necessity

For Smolin, as a working cosmologist, the limitation (understood as a constitutive incompletion) of science that may follow from the suspension of the mathematicorationalist metaphysics operating within science itself is not disabling or disempowering. On the contrary, a temporal naturalism, disabused of metaphysical prejudice, properly allows cosmology to approach the cosmos as a dynamically evolving (albeit nontotalizable) whole and to devise theoretical hypotheses and experimental research programs that would be able to test those hypotheses. One of the most significant barriers to progress in fundamental physics and cosmology, Smolin (2013, 238–39) argues, is the fact that the overestimation of mathematics has led, within areas such as string theory and multiverse theory, to the positing of realities whose purely mathematical and theoretical description can never be experimentally verified as such. The shift beyond horizons of experimental verification is a major barrier to progress within science and is incompatible with the fundamental motivation of scientific enquiry itself. To this extent, temporal naturalism, although its antimetaphysical critique of mathematics may suspend figures of absolute, totalizing, or complete knowledge, will nevertheless give science a greater chance of inventing new testable theories and hypotheses and making progress in the future.

As well as freeing physics and cosmology to invent new research agendas that are unencumbered by inherited metaphysical prejudice, the sense of limitation and incompletion imposed by temporal naturalism perhaps also frees thought up as thought, both in science and in philosophy. The loss of the godlike point of view that accompanies the mathematical determination of supposedly eternal laws governing a harmonious and orderly cosmos is countered by the fact that, when confronted with the radically unknown and even the radically unknowable, thought, both scientific and philosophical, gains greater speculative freedom and perhaps greater potential for imaginative invention or insight. Already in *The Life of the Cosmos* Smolin was arguing that "it is possible, indeed, I would argue necessary, to speculate. Because we cannot invent when we cannot conceive, the construction of a new theory must involve, or perhaps be preceded by, attempts to imagine the outcome" (1997, 5). Purely speculative thought, insofar as it is experienced at or upon a specific yet indeterminate limit, may be a precondition for scientific discovery and for the invention and innovation of scientific theory.[30] As has previously been argued, this is not a limit that divides a (knowable) phenomenal realm from a (forever unknowable) noumenal realm. Rather, it is a limit, itself ungraspable as such, that lies somewhere between what we already know and what we do not yet know. More generally, it is a limit that separates the multiplicity of real entities and structures with which we can come into a real relation from the totality or whole of the cosmos with which we can have no relation at all.

This recognition of the possibility and necessity of a speculative moment within thought is the point at which Unger and Smolin's temporal naturalism and Nancy's cosmology of relational sense can be most closely aligned with each other while maintaining their difference and distinctiveness. Unger's allusion in *The Singular Universe* to the limits of natural philosophy and the powers of science (his affirmation of "our groundlessness—our inability to grasp the ground of being or existence"; Unger and Smolin 2015, 102) repeats, in different terms, the experience of thought and existence in an absence of ground that was described in the context of Nancy's reading of Kant in chapter 1. In *Time Reborn*, Smolin also underscores the limits of scientific knowledge in relation to the question of being as such: "What is the substance of the world? [. . .] What is the essence or existence of a rock? We don't know; it's a mystery that each discovery about atoms, nuclei, quarks, and so on only deepens" (2013,

266). There is much about Smolin's philosophical position that appears to be entirely consistent with a kind of radical empiricism. Science must be stripped of metaphysical trappings in order to liberate empirical practice. In turn, what science comes to tell us about reality must be limited to what is experimentally verifiable such that ultimate questions relating to the "being" of reality itself remain unanswered. Yet it should be noted in this context that Smolin's uncompromising relationism owes much to Leibniz (whom he frequently cites) and that a Leibnizian rationalism founded on the principle of sufficient reason aligns itself in his thinking with both a demand for concrete empirical agendas within scientific practice and a recognition of the necessity of speculation within theoretical and philosophical reflection about science. If anything, then, Smolin's temporal naturalism could be defined as a hybrid thinking—not simply and solely as empiricism, but rather as a form of speculative-rationalist empiricism.

In contrast, Nancy's philosophy of withdrawn sense, despite its emphasis on touch, contact, and the sensory dimensions of being-to, is not in any way an empiricism and arises, as was shown in chapter 1, from a critique of speculative rationality and from an experience of thought rather than from an empirical or scientific experience of the world. Sense is the excessive or withdrawn being of the "there is" (of the world, of nature, of the cosmos) as it is by itself and as such. Nancy's philosophy of sense, then, if it can be aligned with scientific thinking and with naturalism, must necessarily be called a speculative naturalism. Where for Smolin speculative thought is a necessity of scientific reflection that occurs at the limits of current knowledge and as an imaginative act in the service of future theory construction, for Nancy, speculative thought occurs at the limits of thought itself, as the quasi-ontology of singular plural sense. Nancy's quasi-ontology is speculative reflection relating to the excessive, nonsubstantive, nonessential stuff of the world and its opening as world in an absence of ground. It speaks speculatively about what Smolin says science does not know, about the substance of the world and its mystery—a "mystery that each discovery about atoms, nuclei, quarks, and so on only deepens" (2013, 266).

It could thus be argued that Nancy's cosmology of relational sense can be viewed as a speculative supplement to Unger and Smolin's temporal naturalism. Both view the cosmos as a relational structure that *is* only in its singular and plural dynamical becoming and one that has no existence or essence outside of that becoming. Both embrace the necessity of a purely speculative moment or experience of thought that opens up at the limits of both scientific knowledge and theoretical or philosophical reflection. Yet although it can be aligned closely with Unger and Smolin's temporal naturalism and the cosmology they propose, Nancy's thought moves one step further to expose itself to its own excess—that is, to being understood as sense. To this extent, its alignment with temporal naturalism arises as a consequence of its own necessary status as a speculative naturalism.

A Speculative Naturalism?

This idea of speculative naturalism allows for a return to the broader question of the relationship of the Nancean philosophy of sense to science and to scientific thinking that has been the preoccupation of this chapter. Here a brief summary of the different stages of the argument may be helpful. First, regarding the question of objects or things that Nancy's philosophy of sense was shown to describe, a relational world was posited in which the existence of entities was constituted out of purely relational structure and as material differentiation. Beyond any human apprehension or cognition, objects, things, or entities were held to exist as material differentiation, and as such in excess of any correlation with anthropocentric structures of knowledge. To this extent, Nancy's thinking appeared to be similar in key regards to the OSR of Ladyman and Ross and opposed to Harman's object-oriented ontology. Second, and in the context of the question of life and Canguilhem's biological philosophy, sense was shown to constitute the very being of living organisms understood as centers of reference within differentiated relational structures. Sense was then aligned with the information carried by DNA and genetic material in Canguilhem's later thought and brought into relation with the thermodynamic disequilibrium of living structures as described by Nick Lane in *The Vital Question* (2015). This opened up the possibility that Canguilhem's alignment of sense with information could be generalized, beyond living organisms, to the organization of structured matter per se and brought together with Nancy's general ontology of sense. This in turn opened the further possibility of a provisional alignment of the speculative category of sense with the scientific concept of information. Nancy's and Canguilhem's thought, taken together, articulate a natural philosophy of life and a naturalized realism according to which the knowledge biological organisms had of their surrounding world was real and was articulated in real relational contact with that surrounding world. Third and finally, in relation to the question of the cosmos and the necessity of a new cosmology, Nancy's relational vision of the cosmos was aligned with Unger and Smolin's temporal naturalism and its image of a singular dynamically becoming universe. This shared image of the cosmos was stripped of all metaphysical ground and external godlike perspective whose temporal becoming needed to be understood in terms of immanently evolving natural, physical processes rather than transcendent, timeless, and universal laws.

In relation to both the biological thought of Canguilhem and the cosmological thought of Unger and Smolin, each had an irreducibly speculative moment. This speculative moment of scientific thought emerges at the limits of what is known or interpretable by science itself. In the case of Canguilhem, this meant that philosophy was enacted as an art of reading biological knowledge in order to interpret the being of life as sense and the organization of the genetic code as sense inscribed in matter. In the case of Unger and Smolin, this meant thinking speculatively about the universe as a structure evolving naturally in time and in the absence of preexisting eternal laws in order to develop testable cosmological hypotheses and in order to, as it were, preimagine future novel theories. The proximity or alignment of Nancy's ontology of singular plural sense with biological and cosmological thought here has been discerned in the light of their shared descriptions of fundamental reality: the understanding of being as sense, of sense as relational structure, and of relational structure as dynamic, time bound, and as evolving in the absence of any overarching unity, ground, or eternal law. Yet above and beyond their shared descriptions, it is arguable that Nancy's ontology of sense, as speculative, can find its point of suture with scientific thought in the speculative moment that ontological, biological, and cosmological reflection all share. Nancy's ontology unfolds as a quasi-ontology in the experience of philosophy and thought at their constitutive limit. Both Canguilhem and Unger and Smolin develop philosophical reflections at the limits of scientific knowledge where interpretation and speculative hypotheses necessarily impose themselves in the service of theoretical understanding or future invention. This confluence of thinking at the limit or limits allows for a precise definition of something like a speculative naturalism. Speculative naturalism occurs, or articulates itself, at the limits of what we know about nature. In the context of science, these are the limits of current knowledge—the edge of what science can at any given moment tell us about the universe and the relational multiplicities that constitute it. For philosophy, these are the limits marked by the impossibility of securing ontological or metaphysical foundations and therefore of grasping any totality or whole. Speculative naturalism, as an experience of thought "at" the limit of thought, remains always in some relation of close proximity with the understanding of the natural sciences. Yet as speculative, it is not limited or restricted to the current state of "best science." Speculative naturalism unfolds as an experience of thought opening onto the limits of both the philosophical understanding of being in its generality and scientific knowledge of physical reality in its specificity. As speculative thought, however, it also allows itself to be exposed as an excess over those limits, delimiting and rendering them indeterminate as such. There is an experience here both of an irreducible and ungraspable finitude of thinking and of an exposure of thought to that which lies in excess of the limit upon which it always takes place as thought.

If the term "speculative naturalism" can be used to describe thought as it unfolds in the juncture (or point of suture) between Nancean ontology and the scientific thinking elaborated here, then it differentiates itself sharply from Harman's speculative realism, which entirely divorces itself from science. It differs also from what Ray Brassier has termed "critical naturalism" (2014, 112) a thinking inspired by and attributed to the philosophy of Wilfrid Sellars. In Sellars's rationalistic naturalism, Brassier argues, philosophy is not subordinate to science but has a critical function insofar as it regulates or "anatomizes" the categories of the "manifest" and "scientific" images of the world and therefore also plays a legislative role in the revision scientific categories themselves (Brassier 2014, 112). As regulative and legislative, the critical naturalism that Brassier ascribes to Sellars and that he himself endorses remains eminently philosophical and distinct from the operations of philosophy at its limit that are at work in Nancy's thought. As was shown in chapter 1, the experience of freedom that Nancy discerns to be at work in the operation of Kantian rationality radically undermines the image of philosophy as a legislative and foundational practice. Speculative naturalism here would therefore be quite distinct from two of the key branches (antiepistemic and epistemic) of speculative realism identified earlier in this discussion.

It would also reveal itself to be entirely distinct from contemporary scientific metaphysics of the kind espoused by Ladyman and Ross. Ladyman and Ross's OSR is also an ITSR, and information itself is considered within their scientific metaphysics to be the fundamental reality of all that can be said to exist. The ontologically primitive status of information is defended by citing a range of eminent scientists and philosophers of science and is therefore justified on the basis of the authority of science itself in accordance with the PPC. Canguilhem's thought opens up the possibility of identifying "sense" entirely with the concept of information as it is used in the biological sciences. If Nancy's use of the term "sense" could be identified more broadly, and via Canguilhem, with the scientific concept of information as used in both biology and physics, then his ontology could be brought into an even closer proximity with the relational universe described by OSR.

This, however, cannot be the case, and the difference between speculative naturalism and the scientific metaphysics of OSR/ITSR must remain as stark and as irreducible as the hard-nosed scientism of Ladyman and Ross would demand. After all, information remains a scientific concept, albeit one imported from cybernetics and communications theory, and it is used in specific contexts within both biology and physical science and relates to what science can experimentally determine and know about material reality. It is because information is the baseline of what science can determine and know that Ladyman and Ross hold it to be ontologically primordial. According to the PPC, anything that cannot be determined or known according to fundamental physics is an idle wheel or speculative toy within metaphysics that cannot be given the status of real being as such. Scientific metaphysics requires the expulsion of all speculation from philosophical thought. In light of this, Canguilhem's attempt to identify biological sense with the information carried by the genetic code cannot easily be entirely carried over into Nancy's general ontology of sense. The biological code is, like all other physical information, determinable and knowable, whereas Nancean sense is always indeterminate, excessive, and withdrawn from phenomenal appearance as the condition of its opening or coming to presence. Therefore, the information known to science and the "sense" posited by speculative ontology must remain distinct from each other.

Although the two categories are distinct, in a certain sense they remain aligned. It can be said that what science determines as information, speculative quasi-ontology in-determines as sense or being. Information is ontologically primitive for Ladyman and Ross only because they cannot concede that anything that remains unknown or unknowable to science can take the name of being or existence. Yet if scientific thinking admits the possibility, and even the necessity, of speculation, as has been seen to be the case both in Canguilhem and in Unger and Smolin, then there is no reason why thought cannot exceed what science determines as information to name the nonsubstantive and ungrounded stuff of being as sense. So it can be said that where Canguilhem views biological sense as information inscribed in matter, Nancy's understanding of sense can be generalized beyond biological life, but only insofar as it is understood, in excess of the concept of information, as sense exscribed as matter. Both information (in OSR/ITSR) and sense (in Nancy) describe the stuff of being as organized relational structure. Science tells us what can be known of such structures as determinable information. Speculative ontology tells us of the excessive, ungraspable, and withdrawn being of the differentiated matter that constitutes structure as such. Speculative naturalism brings both together by thinking at the limits of what they each can know or determine within thought itself and by thinking the very structure of being *in* and *as* excess.

One might wonder what the affirmation of this speculative moment within ontology and scientific thinking really brings, or adds, to either the philosophical or the scientific understanding of reality. Would it not simply be easier, and indeed far more rigorous, to follow Ladyman and Ross and to restrict metaphysics to what our best scientific theories and knowledge can tell us about reality? Or would it not be more plausible to restrict speculative thought to the interpretation of existing knowledge (Canguilhem) or to specific scientific or cosmological hypotheses which might foster future knowledge (Unger and Smolin)? What is gained from moving further, in a perhaps entirely arbitrary manner, to name being itself and in general as something like "sense"?

The answer can be found in a return to the problems of scientism, eliminativism, and reductionism that arise in the context of contemporary naturalist thinking and in relation to contemporary scientific metaphysics in particular. The scientism of contemporary naturalism and scientific metaphysics tends to deny or eliminate large swaths of qualitative experience that cannot be accounted for by the scientific method. For instance, Ladyman and Ross deny not only the fundamental existence of entities or things but also the existence of qualia and therefore, by implication, the reality of qualitative experience per se (2007, 154). Yet we nevertheless do have such qualitative experience in every moment of our lives. This arguably demonstrates the fundamental limitations of any position that disqualifies the reality of qualia and of qualitative experience simply because they cannot be measured by science. The same problem is faced by positions within philosophy inspired by neuroscience, which are eliminative with regard to thought itself. The temporal naturalism of Unger and

Smolin is at least unequivocal in its rejection of eliminative scientism: "We would like to have a naturalism that does not reduce human experience and aspirations to illusion" (2015, 357). Their thinking represents a significant step in moving contemporary naturalism beyond scientism and eliminativism with regard to qualitative experience (480–81).

As has been already argued in the context of Canguilhem's biological thought, what is important about the notion of sense is that it can refer both to the qualitative relation of contact, touch, and sensing through which a shared world is opened up for living organisms and to the material differentiation that constitutes the being of all entities as such. What is decisive about the notion of sense is that it places lived qualitative experience and material structure or process on the same ontological footing and in a relation of seamless ontological continuity each with the other. Although the latter can be determined or measured as information, the former cannot. However, this is no reason to deny it fundamental ontological reality. Speculative naturalism, as described here, by naming fundamental being as sense, allows for qualitative experience and the material reality measured by science to be brought together and understood as part of a greater structure of purely natural and physical processes without the reduction of any dimension of existence to another and without the elimination of what cannot be quantitatively determined or measured.

Nancy's experimentation of thought at its limit therefore leads us to an ontology of singular plural sense that opens onto a more developed naturalist thinking and a realist account of naturalized knowledge. Yet as naturalism, which can be aligned with specific trajectories of biological thinking in the twentieth-century French tradition and with contemporary emerging trajectories in both biological and cosmological thought, it remains speculative and finds its point of contact with scientific thinking in a shared zone of speculative necessity that opens up at the limits of the known and the knowable. As such, insofar as it does open onto a form of naturalism, Nancy's philosophy of sense must do so as a speculative naturalism. The strength of the speculative naturalism that can be derived from Nancean thought is that it closes the gap between the qualitative experiences that take place in nature, in human, animal, and all other biological life on the one hand, and the material structures and processes of physical existence on the other. Yet it does so without reductivism or eliminativism, and without inscribing itself within a totalizing or unifying horizon of thought and knowledge. The speculative naturalism that can be derived from Nancy's ontology of sense does indeed offer a picture of the whole or totality of things, of life, and of the cosmos, and of all that can be experienced within that whole. Yet this is so only to the extent that the whole be thought as infinitely in excess of itself, a relational structure of ceaseless and groundless spacing, opening, differentiation, and becoming. A whole can never be thought or grasped fully and as such.

3 GENERIC SCIENCE

IF LARUELLIAN "SCIENCE" DESCRIBES the formal structure of the real as a radical immanence and therefore as an indivisible and entirely unknowable One, then it may be very hard indeed to imagine how it can have anything plausible or coherent to say about science understood in more conventional terms as the "natural sciences" of physics, biology, chemistry, and so on. At best it might seem that Laruelle's nonphilosophy would offer an antirealist critique of scientific objectivity and have its outcome in some further variant of constructivism or relativism with which debates in the philosophy of science have been all too familiar. Yet what will have become clear from the presentation of Laruelle's thinking in chapter 1 is that his understanding of science or theory refuses to be assimilated to—and indeed seeks unambiguously to distinguish itself from—philosophy as such and in general. Thus, on Laruelle's own terms at least, a nonphilosophical or scientific critique of the status of the natural sciences will not and cannot simply be another variation on or within the philosophy of science.

What Laruelle's conception of nonphilosophy as science offers is a total reconfiguration of the relation of philosophy to science and a unique and highly innovative understanding of scientific realism. His central insight is that the sciences themselves are nonphilosophical through and through and that they share with nonphilosophy a specific posture with regard to the immanence of the real. It is on the basis of this shared posture that nonphilosophy can be understood as a science and that the natural or empirical sciences can be understood as nonphilosophical. For Laruelle, the problem is that philosophy has, in one way or another, consistently bound itself up with the sciences, and has done so both in its attempts to provide epistemological foundations and in its interpretations of scientific results and theories. Laruelle's contention is that philosophy has consistently sought to give a foundation to scientific knowledge or to ground its "objective" vision of reality. Most obviously it has done this by way of epistemology or within the discourse of philosophy of science. The result is that the image we have of science and of scientific theories has become permeated with philosophical representations and metaphysical prejudices. Laruelle's perspective is quite close here to that of Nancy, Barrau, Smolin, and Unger, who, it was shown in the previous chapter, all called for a separation of science from the inherited metaphysical prejudices that have arguably come to characterize its general image and so many of its most successful theories, particularly those of fundamental physics. Yet he goes a step further insofar as he suggests that we have entirely lost sight of what science really is and how it is radically different from philosophy. In *Theory of Identities*, he argues that "science's essence was 'forgotten' or denigrated by philosophy. Philosophy conflated science with its own operations, with the project or with objectivation; it mistakenly assigned to it the unique task of knowing the object" (Laruelle 1992, 78; 2016, 54). By way of epistemology or of specific configurations within the philosophy of science, we have called on philosophy to give some kind of general or universal authority to science's "objective" knowledge of the real. Despite this, Laruelle argues that "philosophy has never *really* founded science. It has instead projected a possible image on it, the possibility of the project of a science" (1992, 103; 2016, 75). By this argument, all that the various positions within epistemology or philosophy of science have ever given us are images or representations of science and its foundations (or lack thereof in more modern modes of constructivism and relativism) rather than any real foundation as such. Yet these images or representations have little or nothing to do with the actual working of science itself and the means by which it effectively procures its knowledge of the real.

This point is worth considering. It might not be too contentious to concede that even though some of the greatest modern scientists may have been philosophically well read, it is not a priori necessary for science to go hand in hand with philosophy in order to develop and pursue its distinct way of knowing. Indeed, it is arguable that the great majority of scientists are not trained or particularly well informed philosophically and have no need of philosophical knowledge or authority in order to do what they do. It is easy to bring to mind a sample of various possible positions within the philosophy of science and key thinkers associated with these positions: conventionalism (Poincaré), logical positivism or empiricism

(Mach, Neurath, Carnap), critical rationalism (Popper), historical epistemology (Bachelard, Canguilhem), postpositivism (Kuhn), variants of constructivism (Bas van Fraassen, Isabelle Stengers), relativism (later Feyerabend), sociology of science (Latour), and contemporary naturalized scientific metaphysics (Ladyman and Ross), as well as realism in its various guises: analytic-scientific (Putnam), structural (Worral), or pluralist (the Stanford School). Of course many other positions, nuances within positions, and important names could be added to such a brief and schematic list. Without wanting to engage in any kind of general overview of philosophy of science in relation to Laruelle, it might simply be remarked that all the positions mentioned are historically contingent with regard to the trajectory of twentieth-century philosophy and have a status distinct from scientific theories as such. They may have engaged to a greater or lesser extent with the scientific and mathematical developments, revolutions, or practices of their day, and it may be possible for the open-minded scientist or scientifically informed individual to find elements of truth or plausibility in all these positions, however much they come into conflict with or contradict each other. Yet it is not necessary for natural science programs within universities to teach these positions when they train the scientists of the future in the various branches and specialties of scientific knowledge. So despite, as Laruelle would have it, the capture or appropriation of science by philosophy, and despite their evident interactions, science arguably has its own autonomous modes of discovering, theorizing, modeling, experimenting, and hypothesizing—modes that exist independent of their grounding in epistemology or in the philosophy of science. If this is so, it is legitimate to pose the question whether science actually needs philosophy or is in any way guaranteed or grounded by it.

Laruelle's answer to this question is unequivocal. Whatever the interactions between philosophy and science may be, and however much the former may seek to project its own image on the latter, science is not in need of any philosophical foundation or legitimation. Laruelle expresses this clearly in *En tant qu'un* [As one], one of his most accessible introductions to nonphilosophy: "Science [. . .] has no need of a philosophical foundation, it is without foundation because it has a cause: the One, the real-as-Identity; not only an immanent cause, but a *cause-by-way-of-immanence*, the causality of radical immanence itself" (Laruelle 1991, 27). This sentence articulates the essence of Laruelle's description of science, and all his thinking about nonphilosophy, theory, and the natural or empirical sciences is consistent with this understanding of the indivisible One of the real as an "immanent cause." It will be remembered from chapter 1 that the specific structure of philosophy is, for Laruelle, one of a division or splitting of the real into instances of (interior) immanence and (exterior) transcendence and their subsequent mixing and synthesis into a conceptual or representational unity. In so doing, philosophy autolegislates and autopositions itself as universal knowledge of the real understood as Being, world, or existence. Science, however, does not seek to represent the real by means of the operations of division, splitting, and synthesis. It leaves the real untouched and unrepresented but has the real as its cause or determination, or, as Laruelle would say, its cause or determination-in-the-last-instance. Science "does not posit itself in order to be, but is determined in the last instance by the real that is here cause (of) itself rather than an auto-positional subject" (Laruelle 1991, 103). Science has its cause outside of itself; philosophy autopositions itself and produces knowledge of being via its own conceptual-representational operations. Put simply, Laruelle is trying to express the way in which scientific knowledge, in its specific production of phenomena, is determined or caused by the real but does not, in and of itself, articulate these phenomena in terms of the universality of being (i.e., philosophically). He describes this as "the scientific *posture* with regard to the real" (Laruelle 1991, 56). This scientific posture is a direct nonrepresentational relation to the real insofar as it is caused by the real and produced without the mediation of logic or conceptual-philosophical determination.

Such a characterization might immediately provoke a number of questions or skeptical responses. Surely it goes without saying that science does have a representational function, and surely it necessarily does use concepts, both theoretical and at times philosophical. Surely also it seeks to make general or universal claims about being or existence insofar as it uncovers the laws of nature or determines and interrogates the existence of, say, natural kinds. What will become clear in the discussion that follows is that Laruelle is serious in his contention that science, in its essence, is a nonrepresentational, nonmediated relation to, or posture toward, the immanent real and that, unlike philosophy, it is caused by the real without seeking to affect or transform the real in its turn. Everything hangs on his meticulous description of what he terms the "posture of immanence" (1991, 50) proper to science and on the logic of causation or determination-in-the-last-instance that characterizes the relation of the real to science. In the light of this meticulous description, Laruelle argues that what is needed is a thoroughgoing nonepistemological redescription of the sciences. The precise specification of science's posture of immanence will open the way for a thought that is adequate to the real relation that science maintains with the real and that will allow science to be stripped of the philosophical concepts and prejudices that have become bound up with it and ingrained in our understanding and representation of it.

What is at stake here, then, is the task of a radical separation of philosophy and science and an eradication of the philosophical image of science. In *En tant qu'un*, Laruelle speaks of this task: "This non-epistemological re-description of the sciences—of all the sciences—founds their veritative re-evaluation and implies a redistribution of the transcendental and the empirical within the interior of each science. [. . .] The remaining a priori structures of scientific objectivity are deduced from this posture of immanence" (1991, 50). So it is worth being clear from the outset. The nonepistemological redescription of the sciences in Laruelle does not have as its outcome an undermining of the "truth" of science or of scientific objectivity. What results from this redescription is a displacement of the basis of scientific objectivity and a transformation of its status. The (illusory) epistemological foundation of scientific truth and objectivity is displaced from philosophy to what it has in fact always in any case already been—that is to say, its cause in the immanent real. The image of science that emerges from all this is at once strange and remarkably familiar insofar as Laruelle's nonepistemological redescription of the sciences themselves does not really touch or alter their specificity or contents but rather transforms their status in relation to the real. In this context, questions, skeptical or otherwise, relating to the representational status of the sciences, or to the existence of universal laws or natural kinds, are themselves also displaced or rendered redundant.

The aim of this chapter is to interrogate and develop further the implications of Laruellian science for a renewed nonepistemological image of the natural sciences. Throughout, a number of questions will be posed or used as guiding hypotheses. Most obviously, what is the image of science that results from this nonepistemological redescription? How does this renewed image relate to the various positions or debates within the philosophy of science without it in turn becoming just yet another philosophical image, position, or variant? What are the implications of this for specific contemporary debates within science itself, if any? What does Laruelle's redescription mean for our most successful scientific theories such as Einsteinian relativity or QM? And most generally, what does the untangling or separation of science from its philosophical capture or appropriation mean for its relation to other forms of knowledge? If the real is unknowable and unrepresentable, does not science become equal to any other form of knowledge, and does not Laruelle's account, despite its professed realism, descend into relativism? Beginning with a detailed exploration of his nonepistemological account of science in *Theory of Identities*, what follows will explore Laruelle's realism-of-the-last-instance as it develops into his mature conception of generic science in work from 2008 onward. Laruelle's generic science indeed gives us an entirely new image of the natural sciences, one that separates them from the totalizing horizon of metaphysical capture and philosophical closure. In this way, and despite their unquestionable realism and objectivity, a principle of insufficiency and modesty is introduced into scientific discovery and knowledge, and the natural sciences are situated within an open horizon. Laruelle allows us to understand scientific knowledge and practice in terms of an unbounded and plural field of the open sciences rather than science understood as some kind of monolith.

Nonepistemology, Identity, and the Scientific Object

Theory of Identities is arguably Laruelle's single most important book on the question of nonphilosophical science and its redescription of the sciences. Later works such as *Introduction aux sciences génériques* [Introduction to generic sciences] (2008) and the major *Philosophie Non-standard* (2010c) can arguably not be properly understood or fully appreciated without reference to this earlier work. Laruelle's other major works of nonphilosophy, such as *Philosophy and Non-philosophy* (1989, 2013b) and *Principles of Non-philosophy* (1996, 2013c), give more general accounts of nonphilosophical science without the more detailed nonepistemological redescription of the sciences provided in *Theory of Identities* (1989, 99–129; 2013b, 97–128; 1996, 43–93; 2013c, 37–77). The theory of identities named in the title aims to outline in detail "a science that is itself rethought and described in its essence in a new way. This new description is no longer epistemological, i.e., philosophical" (1992, 10; 2016, xviii). According to this theory, the realism-in-the-last-stance of the sciences is a theoretical realism in which the phenomena produced by and received into the domain of scientific knowledge are "caused" by real identities that can never themselves be known, thought, or in any way determined or represented. The "identity" of the real, that is to say its indivisible Oneness and absolute autonomy, or indeed the "identities" of the real in the plural, cannot be reduced to anything that philosophical discourse represents: objects and things (phenomena), transcendental objects (= x) or things-in-themselves (noumena), substances, differences, kinds, laws, structures, and so on. It is on this basis that Laruelle's redescription of the sciences sets itself firmly against "philosophical *normalisations* of science," or against any metaphysics of science, and rather seeks to articulate "its displacement [. . .], on the terrain of science," one that will be irreducible to "any (philosophical) image of science whatsoever" (1992, 31, 29; 2016, 15, 13).

In order to elaborate this theory, Laruelle needs, as he did earlier in *En tant qu'un* (1991), to give a theoretical account of the manner in which science relates to the real that is distinct from that of philosophy. He absolutely rejects Heidegger's claim that science does not think. Laruelle wants to demonstrate that science has its own "authentic and consistent thought," one that is entirely autonomous with respect to the operations of philosophy and to the various modes of articulating the philosophical decision (Laruelle 1992, 33; 2016, 16). In particular, he wants to show that it is above all not a question of simply reinstating a notion of the simple givenness of scientific objectivity by way of a naive (or indeed sophisticated) empiricism or positivism (these being variants of the philosophical decision). As was shown at some length in chapter 1, Laruelle's practice of nonphilosophy in general is an experimental thought that arises from a specific experience of and within thought itself, and that gives rise to guiding hypotheses or initial axioms from which all else follows. This experimental practice or experience of thought lies at the heart of Laruellian science. The simple initial axiom derived from this experience is once again that of the anteriority or precession of the real in relation to philosophy, but also in relation to all phenomenal or empirical experience of world, subjectivity and objectivity, consciousness, and so on. The key thing for Laruelle about science in general (or what he will call "first" and later "generic" science—more about this later) and about what we normally think of as empirical or natural science in particular is that "its theoretical criteria are immanent to it; it does not expect to receive them from philosophy" (1992, 59; 2016, 39). This is simply restating what amounts to the assertion, mentioned above, that science does not need philosophy to do what it does. But rather than a simple assertion, this statement can be taken as a description of the experience of science and of the empirical sciences, a description of their specific relation to the real, or to radical immanence, which constitutively bypasses philosophy.

Another way of putting this might be to say that the objectivity of science is not derived from, nor does it in any way depend on, the philosophical conception or representation of that objectivity. If an experiment produces a certain set of results, such as the simple diffraction of light though a prism, or the vastly more complicated and sophisticated instrumental readings produced from the trajectories and collisions of particles in a high-energy accelerator, then it does so because these results have their cause, in-the-last-instance (more about this later), in the real and not in the conceptual compositions and representations of philosophical understanding relating to scientific objectivity or epistemology. Laruelle puts this simply, saying of science: "Instead of founding science's reality on its objectivity, we found its objectivity on its reality" (1992, 62; 2016, 41). In a way, it really is as simple as this formulation suggests. To repeat, science does not rely on philosophy or on epistemological grounding because it has its cause entirely outside of itself in the indivisible real. In turn, the real is autonomous and entirely its own cause because it does not require human thought or representation to be what is, being prior or antecedent to thought and representation. But how is this conception of science different from the philosophical position of scientific realism that would hold that a human independent reality is known by science? In order to answer this question, Laruelle's understanding of Identity (which he capitalizes), unilaterality, and the logic of causality-, or determination-, in-the-last-instance, needs to be explored in much more detail. On this basis, scientific realism, understood as just one possible position among many within the philosophy of science, can be distinguished from Laruelle's immanent realism or his realism-of-the-last-instance.

Laruelle's theory does not just posit Identities as immanently real and therefore as prior or anterior to all thought or representation (as their cause). This causal anteriority or precession of real Identity over thought and representation is also, and most importantly, one without reciprocity or return. An Identity causes or determines something that might appear as a phenomenon or as an object of thought, but it does so without itself being touched, affected, or in any way caused or determined in return. Laruelle speaks of "the real's unreciprocated precession on philosophy," and in relation to real Identities, he speaks also of their "irrevocable precession on every representation (even the theoretical)" (1992, 24, 41;

2016, 9, 24). This logic of nonreciprocity or nonexchangeability is exactly what distinguishes the theory of Identities, and Laruelle's conception of science more generally, from his conception of philosophy. Philosophy, remember, is always a separation of the real into instances of (external) immanence and (internal) transcendence, followed by their mixing and synthesis in the work of conceptual determination, and it is therefore always an (illusory) affection or transformation of the real by thought. Philosophy seeks to touch on the immanence of the real and to transfigure it into its own conceptual-representational operations. The theory of Identity, and Laruellian science more generally, in its logic of nonreciprocal or one way causation, leaves the real entirely untouched and undetermined.

Laruelle can thus say of his theory of Identities: "Far from being a last mode of ontology's autodissolution, it is founded on the immanence of *real* Identity (to) itself, on its absolute nonconvertibility with Being (philosophy of the Ancients) and with the Other (Philosophy of the Contemporaries)" (1992, 41; 2016, 23). Laruelle does not, like Nancy, need to trace an autodeconstructive fatality of philosophy by means of which it is brought to its point of terminus or outer limit. Laruellian science does not need to encounter any limit or terminus of philosophy because, on its own terms at least, it was never part of philosophy. Founding itself in the immanence of real Identity, Laruellian science knows nothing of the divisions and differences that structure the operations of the philosophical decision and convert the real into so many different versions of sameness and otherness, being or alterity. It leaves all logic of difference aside in favor of the sole logic of indivisible Identity and its nonreciprocal and nonexchangeable causal anteriority in relation to all thought and representation. It is in this sense that Identity or Identities are not some kind of real object posed by thought as unknowable (i.e., a transcendental object = *x*, a thing in itself, an excess or alterity). As Laruelle himself puts it: "Identity is not any object whatever [. . .] but the 'prototype' of every real object" (1992, 79; 2016, 55). Immanently real and prior to all the differences that can allow thought to think any object as such (as knowable/unknowable, as real, transcendental, empirical, for-us, in-itself, etc.), an Identity is only ever a nondetermined, nonreciprocal, or, as Laruelle will say, unilateral cause.

It is now possible to specify what Laruelle understands by Identity in relation to the nonepistemological redescription of the natural sciences more generally and also to begin to specify how the theory of Identities yields an image of science grounded in the logic of unilateral causation or determination-in-the-last-instance. It was noted earlier that the nonepistemological redescription of the sciences does not touch or alter their contents; that is to say, it does not add to or remove anything from any given science but simply (albeit radically) alters its overall status in relation to the real and to philosophical representation. The nonepistemological redescription of any given science or scientific theory will, however, radically transform the way we may seek to add metaphysical assumptions or philosophical representations to that theory. It will allow for the extraction or suspension of metaphysical assumptions or representations that may already be embedded in that theory itself. It will, however, leave the essential structure of the theory, as scientific theory, untouched—its presentation or production of phenomena, its mathematical formalism and predictive possibilities, its theoretical-conceptual form. The Identities theorized by Laruelle, insofar as they are immanent, real, indivisible, and anterior to all thought, to representation, and to phenomenalization as such, are never incorporated into a given science as objects of knowledge in such a way as to change or modify that science. They are posed by science in general as the immanent cause of what any science in particular may produce by way of phenomena, knowledge, and so on. Identities are not incorporated into knowledge as such and are not therefore determined or affected by it. On this point Laruelle is quite clear:

> If thoroughly undivided identities are already given (and, if they have to play the role of the cause-in-the-last-instance, they can only be given unconditionally), then they are not given in nature [. . .] or in philosophy, *but in science itself and as such rather than in its regional "objects" or its philosophical Idea.* They are indeed what we can call "theoretical objects," discovered or given as such; but they cannot in fact be discovered on the plane of constituted knowledge. (1992, 85–86; 2016, 61)

If Identities can be said to be objects of any kind, then they are purely theoretical objects posed as the prototype of any possible object that can be represented by philosophical thought or within the field of scientific knowledge as such. Any possible philosophical objects (phenomena, noumena, etc.) and any possible scientific objects (subatomic particles, atoms, molecules, cells, organisms, etc.) have an Identity as their immanent cause, and this Identity is never and can never be an object of knowledge or a determination within the field of knowledge. It remains only a theoretical Identity posed in and by science as a real immanent cause and has no other status outside of science conceived in this manner.

The theoretical conception of Identity thus serves to give us a different general image of science and to understand differently the production or constitution of "objective" phenomena, data, or results within the sciences without intervening in the processes of that production or constitution as such. Science becomes a theoretical and nonphilosophical knowledge of real, causal Identities, understood as that which cannot be known, rather than a knowledge of the real founded in (epistemologically guaranteed) objectivity. In this way, the global, philosophical horizon of science, and the philosophical image of science as objective knowledge of the real, is suspended: "Identity is what allows us to rectify our idea of science 'in general' and to wrench it away from the philosophical horizon. We will call 'science' the manner of thinking that relates phenomena to their Identity as their cause-of-the-last-instance and does so by means of the theoretical representation (of) this cause. A science is the theoretical knowledge not of phenomena, but of their cause (the Identity of the real) by means or the 'occasion' of these phenomena" (Laruelle 1992, 88–89; 2016, 63). As a "manner of thinking," science is a theoretical means of understanding phenomena, distinct from philosophy, that does not seek to change the field of knowledge that constitutes those phenomena in the first, second, or third instance. It seeks only to relate those phenomena back to the Identity of the immanent and indivisible real as the cause or determination of the very last instance. Here it becomes necessary to say more about Laruelle's logic of causation- or determination-in-the-last-instance (DLI; or *détermination-en-dernière-instance* in French, DDI).

Most obviously, the logic of DLI in Laruelle can be related to his engagements with Marxism and to his nonphilosophical redescription of Marxism.[1] Yet his use of the term long predates its explicit and extended elaboration in relation to Marx in 2000. In *Theory of Identities*, Laruelle gives a precise definition of the logic of DLI, describing it as a "unidirectional or irreversible causality, always exerting itself from the real toward the mixtures of effectivity, from the Determined to the Determinable" (1992, 141; 2016, 108). DLI is thus a kind of generalized causation of the phenomenal world, of all thought and appearing, by the immanent real that precedes them. Anything that can be thought, represented, or determined in the world is caused by the immanent real. Laruelle is careful to note, both here and in other works such as the later *Introduction to Non-Marxism* (2000, 2015) that the causation of the real described by DLI is nothing like the modes of causation known to or described by philosophy. DLI is not a causation that exists between determinable phenomena by way of processes, dynamic structures, events, forces, or action and reaction and so on. Rather, it is a "a new form of causality unknown to metaphysics" (1992, 141; 2016, 108) because it is exerted solely, unilaterality, or unidirectionally and nonreciprocally by the real on all phenomenal appearance, thought, or representation in general. This is made clear in *Introduction to Non-Marxism*: "DLI is a causality that is simultaneously real, universal, immanent, and heterogeneous or critical and, as such, DLI is not included in the four forms of the causality of Being (final, material, formal, efficient)" (Laruelle 2000, 41; 2015, 44).

As a mode of causality-in-the-last-instance only, one that is irreducible to all forms of metaphysical causality or to causal relations between phenomena, DLI does not negate or cancel causes of the first, second, third, or *n*th instance. Whatever the multiple layers of causality that can be determined in relation to a phenomenon, and however complex they may be (in terms of processual dynamics, relations of reaction, retention of past traces, etc.), DLI simply states that in-the-last-instance, the cause of that phenomenon will be an immanently real and indivisible Identity, and that the phenomenon itself will have no reciprocal cause or affection exerted on that real Identity, thus leaving it entirely undetermined. Again, Laruelle gives some helpful indications in this regard in *Introduction to Non-Marxism*: "Every secondary causality, as multiple as it is, is only taken into account and introduced within the final 'reckoning' on the condition of 'passing' through the principle causality [. . .] toward which the secondary causality is by definition 'indebted'" (2000, 40; 2015, 42). It is in this sense that any given phenomenon is an occasion for the theoretical knowledge of the cause-in-the-last-instance of that phenomenon—that is to say, the Identity of the real. There are echoes here of the theological doctrine of occasionalism. This doctrine distinguishes between divine causality on the one hand and natural causality on the other, and affirms that God is the one and only true cause of everything. By the same token, all natural or physical causes are not real causes as such but are at best only occasional causes. The difference between occasionalism and Laruellian science, of course, is that the DLI of the latter poses the real, immanent One, or Identity, as the sole true cause and not the transcendence of a divine Being.

The implications of the logic of DLI are far reaching, particularly so in relation to the empirical or natural sciences in the context of their non-epistemological redescription, for what it does is cut across or simply suspend all the different and possible philosophy of science positions that may be brought to bear on any given scientific theory, experimental practice, or body of knowledge. For instance, Poincaré's conventionalism, insofar as it argues that the axioms of geometry are choices rather than experimental facts or synthetic a priori judgments, will hold that one of the most important things about scientific-mathematical representations of the universe is the type of geometry we choose in order to frame these representations, such as Euclidean (Newtonian dynamics) or non-Euclidean (Minkowski space-time and Einsteinian relativity).[2] Inheriting from Poincaré in different ways, both logical positivism and, say, Bachelardian historical epistemology, would stress that logical and mathematical structures or historically evolving conceptual structures are all decisive in the production and evolution of scientific knowledge. Conversely, scientific realists would emphasize the positive truth and objective reality of the content of scientific knowledge independent of the epistemic conditions (logical, historical, or conceptual) that produce it. Then again, constructivists, such as Latour, would stress the various social and institutional processes that contribute to the production of scientific facts as being decisive factors, thus relativizing scientific truth into a wider field of other possible forms of knowledge production. Without going any further in this schematic sampling of possible philosophy of science positions, it can be argued simply that what separates or differentiates them from each other is their respective ordering of the varying degrees of cause in the production of scientific knowledge. For realism, mind-independent reality "out there" is the principal cause of scientific knowledge, and for contructivist antirealism, it is contingent social processes and norms. For conventionalism and its inheritors in logical positivism and historical epistemology, epistemic dispositions (mathematical, logical, and conceptual) and their relation to experience take on a primary role. The key point here is that Laruelle's logic of DLI would be able in principle to accept all these positions and affirm all their accounts of the various causes of scientific knowledge. However, all the differing causes they seek to order and affirm in one way or another must now be considered to be differing layers of secondary causes that are only taken into account or "introduced within the final 'reckoning'" insofar as they all pass through the principal causality of DLI to which they are all equally indebted. Each of these positions arranges the philosophical decision differently according to different characterizations of the empirical and its transcendental conditions in order to give a philosophical image of science's representation of the real. The logic of DLI both accounts for and suspends all these variations on the philosophical decision, and all their respective representations of the real, in favor of the theoretical knowledge of Identity as an indivisible, immanent, and real cause-of-the-last-instance only—that is to say, as an antecedent, entirely undetermined, unrepresented, and nonreciprocal cause of all other possible determinable causes. The causes of knowledge or epistemic conditions represented by conventionalism, logical positivism, critical rationalism, historical epistemology, postpositivism, realism, instrumentalism, constructivism, and so on are all rendered equal in relation to the Identity of the real. The differing configurations of causality or determining ground that they represent, epistemically, logically or ontologically, are all subjected to the causality-in-the-last-instance of that unrepresented and undetermined immanent real Identity.

Given, as was argued earlier, that many working scientists have no knowledge of, or indeed need for, epistemological or philosophical justifications for what they do, this suspension of philosophy of science positions by the theory of Identity and the logic of DLI might well be a matter of supreme indifference, at least to many scientists themselves. A more important question for the working scientist may relate to the status that the theory of Identity and the logic of DLI confer on the objects of science as well as on results, data, and the hypothetical, conceptual, and theoretical frames that are nevertheless needed to interpret them or to give them meaning. Laruelle (1992, 100; 2016, 73) argues that objectivity, as we normally conceive it, confuses the real that is known with the object of knowledge (that is to say, the object as represented in and by the discourse of knowledge). This is the primary mistake of scientific realism—a mistake that renders it philosophical through and through. It conflates the objects of scientific knowledge and representations of those objects with a positive, "truthful" presentation of mind-independent or human-independent reality out there. Conversely, Laruelle distinguishes between Identity, real objects, and those objects of knowledge or phenomena that are determined and represented within any given field of scientific endeavor. He appears at times to equivocate as to whether science knows or possesses objects as such. So on the one hand he writes: "Science has a *real object*, but this object derives from its cause or from the One, and it *is an object at the same time as One-in-the-last-instance*" (1992, 60; 2016 39). But on the other hand, and as has been repeatedly underlined here, Laruelle also says that the Identity of the real "is not any object whatever," noting further that "rather than an *ob-ject* in the philosophical sense, it is a cause, and the true 'objects' of science are *causes* rather than *objects*" (1992, 79, 81; 2016, 55, 57).

What Laruelle in fact appears to describe is a unilateral or unidirectional line of immanent causality or determination according to which the Identity of the real is always the first and last cause of what is posed by theory as a real object and then subsequently determined within representational discourse as an object of knowledge. A real object is not really an object as such but itself an immanent unilateral cause of whatever philosophy or scientific discourse might finally be able to represent as an object of knowledge or as a phenomenon. This unilateral line of immanent causality might be schematically represented like this:

(immanent real) Identity→(theoretical) real object→object of knowledge

The arrows here are of the utmost importance insofar as they underscore the unidirectionality of this causal line: the anterior terms always determine or cause what follows without themselves being determined, caused, touched, or affected in return. This leads Laruelle to use the language of "reflection," arguing that knowledge reflects the real but, according to a logic of strict noncircularity, reflects nothing back on the real in return. In this context, he speaks of "the noncircular relation, free of reciprocal determination, that makes it so that a knowledge is subject to the real and only claims to 'reflect' or describe it through the very operation of theoretico-experimental production of its representations. It finally founds the *object of knowledge* as the articulation of empirical procedures [. . .] qua phenomenal objective givens and aprioric procedures of the real object" (1992, 92–93; 2016, 67). So scientific activity certainly does produce representations of phenomena out of its theoretical and experimental activity, but as objects of knowledge, these phenomena are only reflections of a theoretical real object and of an immanently real Identity that are never objectified as such. Laruelle formulates this as follows: "Science does not ultimately have an object in this sense or in the ob-jective sense. The known real and the knowledge of this real form an irreversible duality. Knowledge is a reflection (of) the real, but a reflection that neither posits nor objectivates it and that has its ultimate nonsynthetic cause in the real" (Laruelle 1992, 99; 2016, 72). So "reflection" in Laruelle can only ever be understood in terms of this "irreversible duality" that pertains between the real and its reflection in the object of knowledge. It is in no way a specular and reciprocal mirroring of the real object in an object of knowledge and of the latter into the former. That which can never be objectified in any way is reflected into the objectification or production of phenomena that constitute the theoretical and experimental procedures of scientific knowledge, but those procedures reflect nothing back onto the real in return.

This means that the object of knowledge, together with the theories, hypotheses, deductions, or inferences that may be drawn from it by scientific activity, are always open to modification, alteration, or subsequent refutation by way of the causality-in-the-last-instance of the real. At the same time, the undetermined and unknowable Identity of the real will always remain intact and autonomously its own cause beyond knowledge and representation. Scientific objects of knowledge and the theories to which they give rise will always change, be modified, or be entirely superseded, as they are exposed to the force of the real-in-the-last-instance by way of the production of new phenomena out of the theoretical and experimental procedures of scientific endeavor itself. This again allows Laruelle to distinguish science from philosophy understood as knowledge that always co-constitutes, touches, and transforms the real that it tries conceptually to represent as being. It also again allows Laruelle to specify a general image of science and of scientific knowledge as opposed to philosophical knowledge and to speak of "an autonomous thought of scientific knowledge [. . .]: this knowledge must represent the 'real object' in an 'object of knowledge' (finite product of the process), which has *the property of modifying itself without thereby claiming to modify the known real, as is the case in philosophy*. Knowledge is not an attribute or determination of the known real, which is a cause absolutely anterior to its knowledge. Science changes the order of its thoughts without also claiming to change the order of the real" (1992, 96; 2016, 70). A powerful and effective understanding of theory change is implied by this account of scientific knowledge as autonomous (i.e., nonphilosophical account) mode of thought. For if such knowledge is produced via the unilateral causation-in-the-last-instance of real objects and immanent Identities, then it is clear that that knowledge will always be open to further, potentially unlimited, transformation or reconfiguration by the causation of a real that itself always remains undetermined and unknown. The theories, hypotheses, and experimental results can always and will always be modified by a real that they can never themselves modify.

This, then, is the image of the sciences that is yielded by their Laruellian nonepistemological redescription. It is one in which the objects of scientific knowledge are no longer a synthesis or conflation of a known real and an idealized philosophical understanding of objectivity. The objects known to science become Unilateralities—that is to say, nonthetic reflections of a real that itself remains unknown. A Unilaterality here needs to be understood as the object of scientific knowledge placed into the relation of irreversible duality with the real that is its cause in-the-last-instance. It is the scientific object understood as a result of the unidirectional causality that runs from immanent real Identity, through to theoretically posed real object and then resulting in the scientific object itself. So what the nonepistemological redescription of the sciences does is unilateralize or dualize their contents so that they are no longer viewed as positive representations of an objective reality out there (realism) or primarily as the result of possible variations of epistemic dispositions in relation to experiment and experience (conventionalism, constructivism, instrumentalism, etc.). Unilateralities are, Laruelle would say, "non-positional reflections of the real." A nonpositional reflection of the real is "a reality affect, an invincible and nonlocalizable, non-identifiable feeling of reality." In this way Laruelle's nonepistemological redescription of the sciences yields a strong realism-of-the-last-instance: "In-the-last-instance alone, and despite everything, despite the absence of every relation of resemblance or of 'tableau' to an object, science presents itself globally as an index of reality, of undivided 'realist' representation" (1992, 306; 2016, 250).

It may now be worth standing back from all this detailed exposition of the realism-of-the-last-instance of the theory of Identities in order to evaluate its wider implications and importance. It has been argued that the key aim of Laruelle's redescription of the sciences is to free them from their capture by philosophy and metaphysical prejudice and thereby to eradicate the philosophical image of science by means of and from within science itself. This project of stripping science of its metaphysical carapace was shown to be a key aim also in the projects of Nancy/Barrau as well as Unger and Smolin, respectively, in the previous chapter. In Laruelle's case, the eradication of the philosophical image of science hinges on the way in which the theory of Identities and of their resulting Unilateralities unhooks science itself from any claims to positively represent the real and thereby detaches it also from the philosophical horizons of being and totality. It has already been indicated that for Laruelle, "Identity is what allows us to rectify our idea of science 'in general' and to wrench it away from the philosophical horizon" (1992, 88; 2016, 63). The mode of this extraction of science from philosophical horizons is worth specifying further. The key point about the Unilateralities that constitute the objects of knowledge in Laruelle's account of the sciences is that they cannot be integrated into any horizon of unity, of unified being, or of any ontological-conceptual generality (positive-realist, negative-apophatic) whatsoever. They only ever exist within a constitutively "open" and multiple horizon that, a priori from the perspective of science and its "realist posture of the last instance" (1992, 74; 2016, 51), can never be equated with an image of reality or universal being, or a potential for knowledge thereof. This goes far beyond the scope of any simple empiricism or instrumentalism that would limit what we can know of reality to the successes of experimental situations and to what can be observed within them. By detaching the objects of scientific knowledge from philosophical-conceptual determination and placing them into a relation of unilateral DLI by indivisible real Identity, Laruellian science suspends the generalized figure of totality and with that all possibility of speaking philosophically-scientifically about being as being. This would also apply even to a metaphysically skeptical empiricism or pragmatic instrumentalism that, like any other philosophy of science position, has epistemological and ontological representations and assumptions at the core of its specific configuration of the philosophical decision. Laruelle says of this extraction of science from philosophical capture: "The effect of the suspension of the Whole and of the philosophical form in general is that there are now only absolutely dispersed *Unilateralities*, whose *chaos* is not limited by philosophical teleologies" (1992, 180; 2016, 142). In this way, the theory of Identities and the posture of immanence proper to Laruellian science do not substitute or alter the form or contents of the empirical or natural sciences but adds to or supplements them in such a way as to markedly reduce their *philosophical* sufficiency. The theory of Identities "adds itself to [the existing sciences] as the special science, which makes clear that the described phenomena have meaning only through and for science" (Laruelle 1992, 155; 2016, 121). Unilateralized by the theory of Identity into an irreversible duality of their DLI by the immanent real, the objects of scientific knowledge are constituted in a field of dispersed multiple singularity that, while not representing the real, are no less real insofar as they are nonthetic reflections of the real itself.

Proximal Positions: The Veiled Real, Nonunity, and the Dappled Universe

The skeptical reader may not yet be at all convinced by the nonphilosophical image of science and of the sciences that has been outlined here.

Even if Laruelle's theory of Identities, as well as its principles of DLI, immanent causality, nonthetic reflection, irreversible duality, and Unilaterality, can be viewed sympathetically as at least being theoretically rigorous or systematically coherent and consistent, the question may still remain as to whether they sufficiently, or in any way adequately, describe the sciences, and with that the specificities of scientific activity and knowledge. Questions relating to the representational status of science and its use of concepts (theoretical/philosophical) have already been raised and in part dealt with, but they are far from being fully resolved. Perhaps most obviously, the skeptical question concerning Laruelle's image of science extracted from the philosophical horizon of totality (of "the Whole and of the philosophical form"; Laruelle 1992, 180) relates to the question of the status of the laws of nature that science can be said to discover and to concepts such as "natural kinds," the existence of which scientific endeavor might be said to presuppose. For many it might be taken as given that science, and in particular physics, has as its purpose the discovery of the laws of nature. A law, understood as a regularity, a rule of physical necessity that pertains under all conditions and in all places of the universe, necessarily has a real generality or universality that Laruelle's nonepistemological description of science would appear to preclude. And surely all scientists are, in one way or another, in the business of producing generalizations that explain phenomena and that will therefore also have predictive power. A law therefore represents a real and positive structure of the physical universe that pertains everywhere. This notion does not seem to sit well with the Laruellian theoretical description of the objects of scientific knowledge as absolutely dispersed and chaotic Unilateralities. Nor do such objects fit with the concept of natural kinds understood as a natural rather than artificial or human-categorical grouping of entities into specific kinds of objects whose identity can be represented.

In fact, and as anyone familiar with philosophy of science debates will be aware, the question of the status, or even the existence, of laws of nature or of natural kinds is very much an open one.[3] In the previous chapter, the possibility that the laws of nature or fundamental laws of physics might be said to evolve in time (and therefore be at least temporally regional) was raised in relation to the arguments of Unger and Smolin (who, it was noted, are in the good company of the likes of Peirce, Whitehead, Wheeler, and Prigogine and Stengers). It was shown also that Barrau argued in his collaborative essay with Nancy that there was much in modern and contemporary physics that called the very idea of fundamental laws into question. It was also indicated that Ladyman and Ross's contemporary naturalized metaphysics and its accompanying OSR denied the existence of things (on a fundamental ontological level) and argued that the concept of natural kinds becomes entirely redundant in the light of their own conception of real patterns (2007, 294–97). OSR does, however, insist that real patterns may amount to something like fundamental laws and that explanatory and predictive generalizations have an essential role within the practice of the sciences (281–90). This discussion is not the place for an overview of philosophy of science debates relating to laws and natural kinds. It may be of interest, however, to compare Laruelle's nonepistemological sciences to positions within scientific thought or philosophy of science that are in certain key respects similar or proximal. The aim of such a comparison will not be to retroactively confer some kind of more official scientific or philosophical legitimacy on Laruelle's account of the sciences. From the nonphilosophical perspective, the positions that will be discussed may be similar in some ways to nonepistemological science, but like the philosophies of difference that are close in many ways to nonphilosophy, they do not go far enough insofar as they remain philosophical. However, the comparison will serve to show how proximal the Laruellian image of the sciences are to certain specific established positions as well as how it differs from them in its radicalization of the nonphilosophical gesture. This will allow for a more developed picture of the Laruellian image of the sciences to emerge and open the way for a detailed discussion of the relation of the sciences to the more general, or generic, conception of science in his later thinking.

Bernard d'Espagnat: The Veiled Real

Bernard d'Espagnat was a theoretical physicist best known for his work around quantum physics, and in particular for his thinking about its philosophical implications in relation to the nature of reality. He was the author of many books that dealt with the conceptual foundations of QM and their philosophical dimension, and he was perhaps best known for his concept of the veiled real. As will become clear, this concept has striking similarities with Laruelle's conception of the real as an indivisible One or radical immanence. It has similar implications also for the way in which science understands the relation of the real to phenomenal or empirical experience and for the place of science more generally in relation to other areas of knowledge. The concept of the veiled real also bears interesting comparison with the OSR of Ladyman and Ross discussed in the previous chapter. It was noted in that context that structural realism could be traced back to the thought of Henri Poincaré, a figure who is discussed by d'Espagnat and who is a clear influence on him.

D'Espagnat's 2002 work *Traité de physique et de philosophie* [Treatise of physics and philosophy] gives one of the most comprehensive overviews of his conception of the veiled real, and in it he offers some significant discussion of Poincaré, noting approvingly that it was he who was "one of those who underlined in the most energetic manner that experimental findings and the theories which link them together should not be understood as descriptions of an underlying independent reality" (422–23). There is a clear rejection here of a certain kind of strong scientific realism: the objective results and theoretical models of science cannot be simply viewed as positive representations of a self-structured and mind-independent reality or as being identical with such a reality.[4] Elsewhere in his writing, d'Espagnat cites Poincaré's comment from *Science and Hypothesis* when he speaks of "the real objects that nature eternally hides from us" (cited in d'Espagnat 2012, 6). Yet despite this approval of Poincaré's rejection of strong scientific realism and emphasis on the hiddenness of real objects, d'Espagnat is quite clear about the evidence and basis of the success of scientific knowledge and discovery noting that there are "two points that are, for the physicist, simultaneously evident and fundamental: the permanence of equations—which, above and beyond their changing interpretations, preserve their heuristic and predictive value [. . .]—and correlatively, the fact that physics develops an always increasing power of prediction in relation to phenomena" (2002, 287). The permanence of certain fundamental equations of physics (e.g., Maxwell's equations) across theory change was, it will be recalled, one of the key concerns of the structural realism developed by Worral (1989) and pursued further by Ladyman and Ross. Similarly, the predictive success of science is often held up as the linchpin of scientific realism by way of the "no miracles argument" (see chapter 2 and the discussion of Wigner 1960).

So despite his apparent rejection of a strong scientific realism that aligns the positive or objective experimental and theoretical representations of phenomena with an independent reality, d'Espagnat nevertheless shares structuralist realism's focus on the permanency of equations across theory change as well as the decisive emphasis on science's predictive success that is so important to scientific realism in general. Yet d'Espagnat's interpretation of the consistency or permanency of the equations of fundamental physics is rather different from that of OSR. Rather than leading him to suppose that the fundamental "real patterns" that can be determined via the mapping of structural relations mathematically or as information (ITSR) simply are the real (a tendency that leads Ladyman and Ross toward Pythagoreanism or mathematical Platonism), d'Espagnat suggests that the mathematical equations that have such permanence and predictive value for science may be reflections of the real insofar as they are caused by it. In relation to the constancy of Maxwell's equations, he argues that it may result more from the fact that the real is structured but not itself mathematical or informational. Rather than endorsing any kind of Pythagoreanism, he suggests

> as more plausible, the idea that the "real"—independent reality [. . .] is structured and that a little of this structure passes into our "laws." In other words [. . .] I hold as valid the notion of an "enlarged causality" that is exercised not from phenomenon to phenomenon, but on phenomena by the "real" [. . .] this causality [. . .] can accommodate that of structural causes and these in turn can be understood within such an approach rather than as simple regularities observed between phenomena. In fact, these "enlarged structural causes" [. . .] are, quite simply, the structures of the "real"; it has been shown that, from my perspective, they constitute the ultimate explanation of the fact that laws—or, in other words, physics—exist as such. (2002, 517–18)

This notion of an "enlarged causality" that does not operate between phenomena but rather is exerted by the real on phenomena lies at the heart of d'Espagnat's notion of the veiled real. Science can observe lawlike regularities that will yield predictive success, not because the observed phenomena or their mathematical or informational determination are the real objectively, but because these lawlike regularities are caused by a greater structured real that remains hidden from all phenomenal manifestation or possibility of scientific determination and without any correlation to it.

This places d'Espagnat's thinking of the veiled real in a relation of both proximity and distance with regard to structural realism, and he is quite explicit about the differences between them: "There is an important difference between the conception of the veiled real and structural realism such as it is generally presented. The conception of the veiled real [. . .] uniquely includes the conjecture that our great mathematical laws would be grossly deformed reflections—or traces that cannot be deciphered into certainties—of the great structures of the 'real'" (2002, 518–19). This perspective sharply differentiates itself from Pythagoreanism or mathematical Platonism, which would argue that the being of the real simply is mathematical. Instead, it proposes a more modest place for mathematics as a kind of approximate or partial modeling of a structural real that always lies beyond what we can know or what mathematics or information can determine. D'Espagnat's view of mathematics would therefore appear to be close to the kind of model proposed by Unger and Smolin that was discussed in the previous chapter. This is borne out in particular in his interpretation of quantum formalism, of QM, and of the so-called measurement problem. The quantum measurement problem turns around the fact that experimentation at the microscopic level can yield only partial information in relation to the state of a quantum entity (either position or momentum but not both simultaneously) and cannot properly account for the shift between the continuous and unified development of the wave function (as described by the Schrödinger equations) and the discontinuous collapse of the wave function. This problem has preoccupied much of the scientific and philosophical debate about QM and has come to be central in the debate relating to the possible existence of multiple universes and the status of the Everett interpretation of QM. For his part, d'Espagnat argues that the predictive power of QM, when taken together with its constitutive limits, uncertainties, or incompleteness, gives ample grounds for his hypothesis of the veiled real and for his rejection of a mathematical vision of the real: "Quantum formalism, being only, in essence, a predictor of observations, cannot be understood as a true grasping of the real, and this is so whether the nature of this 'real' is purely mathematical or not. Pythagoreanism is an inspiring vision but it in no way provides the assurance of an access to the real in itself" (2002, 497). At the most fundamental level, and by means of our most powerful and predictive fundamental scientific theory (QM), we do not seize or capture the real in itself.

This interpretation of the quantum measurement problem as limiting our access to the real in itself is not far from an instrumentalist account, which also argues that what we can know of reality is limited to the success of experimental situations and what can be observed within them. Where d'Espagnat's conception comes much closer to Laruelle's axiomatic understanding of the real as an indivisible, immanent One is in his notion of an "enlarged causality" pertaining between real and the phenomenal or empirically observable world. Enlarged causality as a concept appears to be close to Laruelle's logic of causality, or DLI by the real. In both instances, it is not a matter of any kind of causality pertaining between phenomena. Like Laruelle's immanent One, d'Espagnat's veiled real is entirely anterior to all phenomenal manifestation as its generalized cause. This means that it is anterior to all of the theoretical conceptual pictures of reality offered to us by modern science and is irreducible to them. So, for instance, he insists that the space-time projected by Einsteinian relativity by way of non-Euclidean geometries is not a deep structure of the human-independent reality of the universe but a property of our humanly perceived phenomenal world and therefore at best a partial or even "grossly deformed" reflection of the "great structures of the real": "an essential trait of the conception of the veiled real is precisely that the said 'real' is conceived as being primary with regard to space-time. That it is not a real in the sense of a thing one can touch but, entirely at the other end of the spectrum, a real in the sense of being" (2002, 504). Clearly Laruelle, as a matter of nonphilosophical rigor and probity, would never equate the real with being, as d'Espagnat does here, but the idea that any projection of space-time is necessarily "not noumenal but phenomenal in nature" (d'Espagnat 2002, 275) is entirely consistent with the nonphilosophical understanding of radical immanence as resistant to all operations of transcendence, phenomenalization, and conceptual determination or representation.

Indeed, d'Espagnat goes further, even more closely echoing Laruelle's critique of conceptual determination as transcendence and as an operation of division and splitting. Not only is the veiled real anterior to the human projections of space-time but it is also prior to any division of spirit and matter, mind and body, subject and object. As he puts it: "According to the conception of the veiled real, the real is primary with regard to the matter–mind division" (2002, 518). This implies a model, similar to Laruelle's, according to which thought, consciousness, perception, or any kind of conceptual determination separates the real from itself (Laruelle would say divides it) in order to present or represent it to a subject or subjective consciousness by way of an object. Where Laruelle axiomatically posits the indivisibility of the real, d'Espagnat posits its nonseparability and therefore its nonobjectifiability, its anteriority with respect to any division into subject and object: "According to the conception of the veiled real, in accordance with non-separability, the 'real' is not in this way 'separable' by thought [. . .] the 'real' is not even conceptualizable for us" (2002, 520).

This formulation is vanishingly close to Laruelle's conception of the real as autonomous indivisible and radical immanence. For both the nonconceptualizable real is not some kind of other world or beyond-world. It is not an elsewhere or transcendent world; it is not an excess or alterity. Rather, it is simply an immanence within phenomena that causes phenomena as such. Laruelle makes this clear when he refers in *Theory of Identities* to the "radical immanence of the One, which is neither a Beyond nor even an Other of the World, but the cause-of-the-last-instance that enjoys its precession on the world. Instead, it is phenomena and, in general, 'objective givens' that are 'beyond' the Identity (of) the real" (1992, 91; 2016, 65). As a mode of transcendence, or separation by way of thought, it is, in a startling inversion, the phenomena that present themselves to us that are a "beyond" of the real rather than the real being a beyond of phenomena, particularly when taken in the light of the unidirectional nature of Laruelle's DLI and d'Espagnat's enlarged causality.

At this point, of course, it may be worth underlining that from the Laruellian nonphilosophical perspective, d'Espagnat's thinking nevertheless remains philosophical despite its close proximity to the axiomatic postulates of nonepistemological science. The principal point here is that in the absence of Laruelle's special kind of axiomatic thinking and method, d'Espagnat's concept of the veiled remains representational insofar as it is conceptualizing the real as veiled. This is consistent with the fact that he can equate the hidden real with that most philosophical of concepts, "being," and also that he can ascribe the quality of structure to the real even while affirming the radical unknowability of that structure. The difference here may be small, but it would be enough for Laruelle to place d'Espagnat's thinking on the side of philosophy rather than on the side of science as he strictly understands it. It needs to be emphasized again that highlighting the proximity of d'Espagnat to Laruelle is not intended to confer the authority of a renowned and respected scientist and philosopher on the nonphilosophical project and its nonepistemological redescription of the sciences. This is so precisely because the thinking of the veiled real remains philosophical. The authority of Laruelle's science, if it has any, comes from it immanently, or from its posture of immanence with regard to the real and the specific experience or experimentation of thought that such a posture makes possible.

What the comparison between the two makes clear, however, is that the image of science that emerges from Laruelle's nonepistemology and the axioms of immanence, indivisibility, and unidirectional causality are not at all incompatible with important philosophical interpretations of, say, QM, except perhaps that such philosophical interpretations do not quite go far enough from the nonphilosophical standpoint. Interestingly, the implications of d'Espagnat's thesis of the veiled real for the relation of science to other areas of knowledge can also be compared to those of the Laruellian nonphilosophical project more generally. At stake here ultimately, and as is argued throughout this book, is an image of the natural sciences that inscribes a greater modesty or a principle of irreducible insufficiency at its very heart, and thus reorientates its relation to other ways of knowing. In particular it renders obsolete the epistemological and ontological excesses of scientism. D'Espagnat articulates this modesty in the following terms: "There is ample reason for the existence of a 'real,' towards whose structures the human mind can only tend, to appear justified [. . .] even though, all the while, there is a consciousness that the human mind will never reach these structures because any intelligence of them by far exceeds its capacities" (2002, 528). This real is not so much a world of noumenal or transcendental objects or things-in-themselves because, as was indicated above, d'Espagnat's real (like Laruelle's) is anterior to any possibility of objectification or splitting into subject–object relations. At the same time, the radical unknowability of this real is not incompatible with the effectivity and permanence of the fundamental equations of physics or with their predictive power.

D'Espagnat's is a realism that is different from mainstream scientific realism insofar as it holds the real to be radically unknowable but affirms the effectivity and the realness of the products of scientific knowledge (by way of enlarged causality). It nevertheless firmly pits itself against the scientism that can so often accompany philosophical attempts to defend a strong scientific realism and thereby elevate the authority of "objective" science over all other forms of "less objective" knowing. It will be recalled from the discussion in the previous chapter that Ladyman and Ross's OSR was also in the service of an explicit defense of scientism, holding that the knowledge produced by physics had to be considered as fundamental, and that all the special sciences—indeed, all other forms of knowledge in general—were required to be bound to that first order of knowledge according to the PPC. Conversely, d'Espagnat's realism, insofar as it takes the real to be radically unknowable, introduces a horizon of insufficiency into scientific knowledge that relativizes its primacy in relation to other ways of knowing or producing knowledge. D'Espagnat expresses this in the following terms: "My conception is, definitively, one of a 'real,' structured certainly, in relation to which I do not exclude the possibility that poetry, the arts, or mysticism can give us some glimmers, but which is no less fundamentally nonconceptualizable by human beings" (d'Espagnat 2002, 519). As was indicated in preliminary terms earlier, Laruelle's theorization of science as opposed to philosophy has, together with his redescription of the sciences, the aim of reformulating or reforming the field of human knowledge and the way in which different regions of knowledge are hierarchized in relation to each other. This is one of the consequences of d'Espagnat's understanding of the real as veiled, bringing his thought once again into a relation of close proximity with that of Laruelle. Yet where Laruelle's is a nonphilosophical realism-of-the-last-instance, d'Espagnat's is a kind of minimal or "open" realism: "I have found it appropriate," he writes, "to adopt as a point of departure for my reasoning the postulate of a minimal realism, that of an open realism" (2002, 273).

D'Espagnat's open realism has strong similarities to the concept of "open sciences" that will be developed below. For now, it is worth noting that Laruelle's realism-of-the-last-instance and d'Espagnat's open realism both open out onto a radical pluralism of thought that is not reducible to any kind of conventional relativism that has become the hallmark of certain strands of recent and contemporary philosophy of science associated with constructivism or antirealism. Indeed, if Laruelle's nonepistemological redescription of the sciences can be brought into a partial alignment with positions within the philosophy of science proper, it is consistently with those positions that, like d'Espagnat, marry a kind of more modest realism with a strong pluralism.

John Dupré: The Nonunity of Science

The idea of the nonunity of science has been strongly associated with what has come to be known as the Stanford School, which includes the two figures engaged with here, John Dupré and Nancy Cartwright, but also other prominent philosophers of science such as Ian Hacking (engaged with in the previous chapter), Peter Gallison, and Patrick Suppes. Broadly speaking, they share a commitment to antiessentialism, antideterminism, and antireductionism while melding a vision of ontological plurality with realism. Again, Ladyman and Ross's ontic structuralist realism comes into contrastive view insofar as one of its aims is specifically to secure the unity of science in such a way that it can be placed into direct philosophical opposition with the disunity theses of the Stanford School. In *Every Thing Must Go*, Ladyman and Ross go so far as to argue that the very goal of philosophy of science, and in particular of a radically naturalized metaphysics, is to ensure the unity of science in general and to unify all realms of the fundamental and special sciences with physics posed as the most fundamental order of knowledge and given a place of supreme authority according to the PPC (2007, 1). Ladyman and Ross take specific issue with the disunity theses of John Dupré and Nancy Cartwright that will be engaged with here (5–6).

For his part, John Dupré, in works such as *The Disorder of Things: The Metaphysical Foundations of the Disunity of Science* (1993), develops a twofold thesis that is at once epistemological and ontological. The first (epistemological) thesis is that of the disunity of science proper—that is to say, "the denial that science constitutes, or ever could come to constitute, a single unified project." The second (ontological) thesis is "an assertion of the extreme diversity of the contents of the world. There are countless kinds of things" (1993, 1). Needless to say, from the Laruellian perspective, these theses are philosophical through and through, and to such an extent that they appear even further away from the axiomatic thinking of radical immanence than d'Espagnat's notion of the veiled real. Nevertheless, the disunity of science theses defended by Dupré do bear some comparison with Laruelle's thinking of scientific objects as Unilateralities. It was shown in some detail earlier that the status of the scientific object was transformed in Laruelle's nonepistemology into a Unilaterality, a nonthetic, nonpositional, and unidirectional reflection of the immanent real that leaves the real itself radically undetermined. Unilateralities, it will be recalled, are not subsumable into any philosophical horizon of unity or totality and are "absolutely dispersed *Unilateralities*, whose *chaos* is not limited by philosophical teleologies" (Laruelle 1992, 180; 2016, 142). In this way, the suspension of the philosophical horizon of totality is central to Laruellian science and to his separation of science from philosophy in general. Laruelle's nonepistemological Unilateralities therefore offer up a picture of the sciences and of the objects of scientific knowledge as being radically nonunified in a manner that has a strong resemblance to the epistemological and ontological theses of Dupré.

Dupré, like Nancy and Barrau, and Unger and Smolin, wishes to separate scientific endeavor and knowledge from its inherited metaphysical assumptions in a manner that parallels Laruelle's attempt to separate science from philosophy in general and thereby to eradicate the philosophical image of science. It was argued in the previous chapter that one of the key issues at stake here was the traditional image of the cosmos as an orderly and harmonious system forming a unified whole. This was the image that Nancy, in *The Sense of the World*, sought to challenge in order to move beyond "the old cosmo-theo-ontology" and to develop an acosmology of relational sense (1993c, 62; 1997, 38). It was also the image that Smolin, along with Unger, sought to overturn in their attempt to think a fundamental order of cosmological time and a dynamic evolution of the universe and its lawlike regularities. Dupré's position echoes that of the scientists Barrau and Smolin insofar as he argues that the traditional image of the cosmos as an orderly and harmonious system does not at all fit with what modern science actually tells us: "I claim that the founding metaphysical assumptions of modern Western science, most notably those that contribute to the picture of a profoundly orderly universe, have been shown, in large part by the results of that very science, to be untenable and this, in turn, shows the impossibility of a unified science" (1993, 2). There is a strong sense in which Dupré's two theses, the first epistemological and concerning the disunity of science, the second ontological and concerning the disunity of the cosmos, are strictly interrelated or interdependent. There can be no unified science because science itself shows the universe to be nonunified—or as Dupré himself puts it, "The disunity of science [. . .] reflects accurately the underlying ontological complexity of the world, the disorder of things" (7). As with Unger and Smolin's position, this is of course a no doubt disputable and contestable perspective, both from the point of view of science and from that of debates within contemporary philosophy of science. As has been indicated, Ladyman and Ross would be among the first to contest this vision of disunity.[5] Yet it is clearly an established and respectable position within recent philosophy of science and one that, like Nancy's acosmology of relational sense, has an affirmation of ontological plurality at its core.

The important point to note here is that Dupré's is a vision of the universe as disordered and as ontologically plural, and that this in turn limits the power of scientific knowledge and pluralizes or multiplies possible epistemic positions without being any less realist and without reverting to relativism or constructivism. So on the limitation of the scope of science, Dupré argues: "Skepticism about cosmic order should raise doubts not only about the unity of science but also about its universality. Since science does, presumably, presuppose some kind preexisting order in the phenomena it attempts to describe, limits to the prevalence of order may entail limits to the applicability of science. Some areas of science may fail because the subject matter is inhospitable to scientific method" (1993, 11). On the pluralizing or multiplication of epistemic dispositions, he notes: "My thesis is that there are countless legitimate, objectively grounded, ways of classifying objects in the world. And these may often cross-classify one another in indefinitely complex ways" (18). So if there are countless and indefinitely complex ways of classifying and therefore knowing objects, this is not because our different ways of constructing models are all of equal value no matter what, but rather because the limitless multiplicity of the real leaves it open to be approached or known in a potentially limitless variety of ways. This is not a relativism but a pluralist realism where the multiplicity of the real is the cause of the multiplicity of epistemic possibilities. At the same time, this multiplicity may set limits on what science may do or know, meaning that it is not only disunified in itself but also severed from any possibility of achieving total knowledge of reality. Dupré refers to this perspective as a "metaphysics of radical ontological pluralism," as "promiscuous realism," or more assertively as a "truly promiscuous realism" (18, 36).

The outcome of this vision is once again strikingly similar to the range of positions that I have surveyed in this volume. There is a separation of science from its inherited metaphysical assumptions, a radical pluralization of the objects of scientific knowledge outside of any horizon of unity or totality, and (in common with Nancy and Barrau, Unger and Smolin, and d'Espagnat) a strong skepticism with regard to the ontological status of mathematics and of Pythagoreanism within physics: "The omnipresent neo-Pythagoreanism of contemporary science is surely not adequately justified by its empirical successes [. . .]. [It amounts to] some kind of commitment to a universe amenable to one orderly and systematic description; a universe in the existence of which [. . .] we have every reason to disbelieve" (Dupré 1993, 224). Once again, this introduction of a certain modesty or principle of insufficiency into scientific knowledge results in a reformulation or reformation of the field of knowledge as a whole, one that rejects scientism and the hegemony of the natural sciences over other ways of knowing. On this point, Dupré is explicit: "No sharp distinction between science and lesser forms of knowledge production can survive this reconception of epistemic merit. It might be fairly said that with the disunity of science comes a kind of disunity of knowledge" (243).

So far, then, it has been shown that there are striking similarities and convergences between d'Espagnat's open realism, Dupré's promiscuous realism, and Laruelle's nonphilosophical realism-of-the-last-instance. It has been emphasized that the likes of d'Espagnat and Dupré are not being called on to legitimate Laruelle's nonepistemological image of the sciences either philosophically or scientifically because nonphilosophy, science, or theory in Laruellian terms only have any force by way of their posture of immanence and as a specific experience of thought. Yet open realism and promiscuous realism as described here do serve to give some preliminary picture, albeit still philosophical, of what the Laruellian non-epistemological image of the sciences will actually look like since they share so many of the same aims: (1) the separation of science from long-standing philosophical-metaphysical assumptions and from horizons of unity and totality; (2) the concomitant renunciation of any ambition on the part of science to know the real in any fully positive or exhaustive manner; and (3) the introduction of an openness or multiplicity into the practice of the sciences that resituates the knowledge they produce in relation to other forms of knowledge without being any less "realist" for all that, and without succumbing to relativism or constructivism. Insofar as Laruellian science and nonphilosophy stand outside of the philosophical decision and outside the philosophical structure of both epistemology and ontology, the image of the sciences he gives will also have to be shown to be different from d'Espagnat's open realism of the veiled real and Dupré's promiscuous realism of the disunity of science. Showing this will be my next task, after I first provide an image of how a realist pluralist philosophy of science can shed light on Laruelle's thinking by way of the work of Nancy Cartwright, in particular in her critique of the concept of laws of nature.

Nancy Cartwright: The Dappled Universe

Dupré's claim that "the disunity of science [. . .] reflects accurately the underlying ontological complexity of the world, the disorder of things" (1993, 7) may still seem surprising. The idea that the success of physics over the last few centuries from Galileo and Newton onward has been founded on the discovery of lawlike regularities is hard to overcome, not least because these discoveries have given us the power to manipulate nature in manifold ways and to make predictions in relation to hitherto unknown phenomena. In the previous chapter I showed Barrau to have well expressed this view, noting that the success of physics seems to be inseparable from its ability to synthesize "the chimeric diversity of the real under a reduced number [. . .] of principles" (Nancy and Barrau 2011, 63; 2014, 34). Yet this view has been repeatedly challenged by all the thinkers discussed here. For her part, Nancy Cartwright places the critique of the notion of constant and universal laws of nature at the very heart of her thinking in works such as *The Dappled Universe: A Study of the Boundaries of Science* (1999).

Where Dupré argues in favor of an ontologically pluralist promiscuous realism, Cartwright pitches her argument in a slightly different way by questioning the manner in which science tends to move from the positing of universal laws to the creation of all-embracing physical theories, and from there to affirming broader metaphysical doctrines about being. So while she endorses a skepticism vis-à-vis the possibility of a unifying metaphysics underpinning all of scientific knowledge, she nevertheless appears happy to remain within a metaphysical register albeit provisionally. Strongly echoing Dupré, her argument turns around the notion that the "dappled" or irreducibly plural nature of the physical universe necessarily limits the applicability of scientific theories to localized or enclosed regions, thereby reducing their power of universality: "Theories are successful where they are successful, and that's that. If we insist on turning this into a metaphysical doctrine I suppose it will look like a metaphysical pluralism" (Cartwright 1999, 31). So if the activity of science is nomological, that is to say relating to the creation of laws, it is so only in a limited fashion to the extent that the regularities it engages with are produced out of local settings and specific configurations of experimental practice that allow those regularities to be measured or framed as such. Cartwright refers to this position as "metaphysical nomological pluralism" and defines it as "the doctrine that nature is governed in different domains by different systems of laws not necessarily related to each other in any systematic or uniform way; by a patchwork of laws" (31). This position appears to concede that something like the laws of nature do pertain, but do so only in a nonunified, nonsystematic manner. In fact, and despite the use of the term "metaphysical nomological pluralism," Cartwright does not think that laws are fundamental aspects of the physical universe. Under certain specific circumstances, we can uncover lawlike regularities and construct laws from those regularities, but underpinning them are what Cartwright calls "capacities," arguing that "it is capacities that are basic, and laws of nature obtain—to the extent that they do obtain—on account of the capacities" (49). There may be some echoes here of d'Espagnat's argument that what we take to be the laws of nature within physics are simply partial and deformed reflections of "the great structures of the 'real'" (2002, 518–19). That is to say, what we take to be laws are caused by some structural force or potential (a capacity precisely) of the underlying real that is not identical with or reducible to the idea of a universal law as such.

So what science does, according to Cartwright, is to set up experimental situations or configurations in which the capacities of nature can be demonstrated to cause regular behaviors or predictable outcomes, and from this laws and theories are constructed. These situations or configurations are what Cartwright calls "nomological machines," and, she argues, it is only by way of such machines that we can "get a law of nature" (1999, 49). Cartwright's argument here resembles somewhat that which has been made by Lee Smolin in relation to the Newtonian paradigm discussed in the previous chapter. For Smolin, the Newtonian paradigm was the method by which scientific experiment isolates subsystems of the universe and concentrates its attention on a limited number of variables, objects, or particles within that subsystem. Smolin emphasizes that although the isolation of such a subsystem would allow for the determination of lawlike regularities, as an artificial construction, it could only ever be "an approximation to a richer reality" (2013, 39) (recalling d'Espagnat's point that our laws may only be a deformed reflection of a great unknowable real). Cartwright defines a nomological machine as "a fixed (enough) arrangement of components, or factors, with stable (enough) capacities that in the right sort of stable (enough) environment will, with repeated operation, give rise to the kind of regular behaviour that we represent in our scientific laws" (1999, 50). A nomological machine is therefore any construction that allows for such fixing of components and factors that in turn allows stable capacities to be effectively determined in their behavior in a manner that is consistently repeatable. Above all, Cartwright notes, "we take experimental design to be a design for a nomological machine" (88). This appears close to Smolin's account of the way experimental situations isolate subsystems of the universe to determine lawlike regularities and forms the decisive moment of Cartwright's critique of the concept of law itself.

As indicated above, Cartwright's thinking arguably articulates a realism that echoes that of d'Espagnat (and therefore also Laruelle) insofar as it is the capacities of the real that, in their effectivity, cause the consistently repeatable regularities that are produced by the experimental configurations of a nomological machine. But at the same time, such a machine needs to be first designed and constructed. Laws are then inferred from or represented out of the results produced from that construction. One might say, echoing Laruelle, that the laws produced within the construct of a nomological machine have capacities of the real as their cause-in-the-last-instance. This may indicate a position that contains both realist and instrumentalist/constructivist elements insofar as capacities are human-independent dimensions of the physical universe, and nomological machines, together with the laws that they yield, are very much human constructions. In any case, in the light of such thinking, the status of laws of nature is radically rethought. The notion of law is stripped of both its universality and its fundamental status. Laws, Cartwright argues, "whether causal or associational, probabilistic or deterministic, are transitory and epiphenomenal. They arise from—and exist only relative to—a nomological machine" (1999, 121). This also places a limitation on the predictive possibilities of science since laws, "including probabilistic laws, obtain only in special circumstances. Concomitantly, rational prediction is equally limited" (158).

What Cartwright's thinking does is to introduce a break in that movement, within the activity of science and of scientific thought more generally, that progresses from the positing of universal laws, to the development of theories, and then to the construction of all-embracing metaphysical narratives. If the universality and fundamental status of laws are questioned or critiqued in this way, then the theories of physics and the metaphysical inferences that can be drawn from (or built on) them also undergo a change in status in relation to the real. If "laws hold only relative to the chance set ups that generate them" (1999, 176), then the theories constructed out of those laws themselves will relate only to those chance set ups (nomological machines) and not to the cosmos as a whole. In this context, Cartwright, like the other thinkers discussed here and in the previous chapter, argues for a limitation of the scope of scientific knowledge in general and its power of totalization, exhaustive knowledge and ontological or metaphysical closure. She does this by drawing on the important distinction between theory and model within scientific practice. Theories in physics, she notes, "do not generally represent what happens in the world; only models represent in this way and the models that do so are not already part of any theory" (180). If theories are produced out of the generation of lawlike regularities by nomological machines, then models, according to Cartwright, have more direct and autonomous relations to the physical world. This allows them to map limited worldlike situations (as described by nomological machines) without being implicated in that movement from law to theory to metaphysical inference that is being called into question here. In Laruellian terms, theories are too much implicated within the structure of the philosophical decision; conversely, models more clearly articulate the posture of immanence that is proper to science as such.

Once again, what we have in Cartwright's metaphysical nomological pluralism is a philosophy of science position that is realist (based on the effectivity of capacities to make themselves known and determined) but that at the same time introduces a radical openness or impossibility of closure into the project of scientific knowledge overall. The fact that this position is realist and gives a strong account of the emergence of lawlike regularities means that it can account both for the permanency of the fundamental equations of physics and for their predictive power (a key starting point for d'Espagnat and for structural realism). Yet the success of physics relies on a different configuration of, or relation between, lawlike regularities, theories, and models. This different configuration or relation means that all-embracing metaphysical accounts of the reality of the universe are not and cannot be the business of natural science. As Cartwright puts it: "Our best and most powerful deductive sciences seem to support only a very limited kind of closure: so long as the only relevant factors at work are ones that can be appropriately modelled by the theory, the theory can produce exact and precise predictions" (1999, 188). By this account, the most important or effective elements of scientific inquiry and knowledge would be the relation between nomological machines and models with theories performing a kind of heuristic mediation between the two.

What Cartwright, Dupré, d'Espagnat, Laruelle, and the thinkers treated sympathetically in the previous chapter all have in common is a desire to separate science from the philosophical figures of totality that permeate it when it becomes blended with metaphysical assumptions. These assumptions are embedded within the implicit or explicit Pythagoreanism of much scientific theory, within its claims to determine the universal, atemporal laws of physical reality and thereby give a unified picture of reality which exists within, or strives toward, a horizon of totality. None of the thinkers here questions the basis of scientific knowledge within the real, and to that extent all are realists. They can account for science's ability to produce knowledge out of the real according to the operations of mathematics and for its incredible predictive power (and without relying on the "miracles argument"). Yet they all, in one way or another, introduce a principle of insufficiency into scientific knowledge that forms the basis of the eclipse of the horizon of totality and any possibility of a unified scientific metaphysics. Cartwright notes, "We may have all the confidence in the world in the predictions of our theory about situations to which our models clearly apply—like carefully controlled laboratory experiments which we build to fit our models as closely as possible. But that says nothing one way or another about how much of the world our stock models can represent" (1999, 209–10).

As has been argued, what these proximal positions within the philosophy of science do is give a more clearly defined image of what exactly the Laruellian nonepistemologically described sciences might look like. Insofar as d'Espagnat, Dupré, and Cartwright are realists, and insofar as they make the real the cause of scientific knowledge (veiled structures, ontological complexity/multiplicity, fundamental capacities), they echo or come into varying degrees of proximity with Laruelle's thinking of knowledge as DLI by the immanent real. Insofar as they strip science of metaphysical assumptions regarding ontological and epistemological unity, universal laws, and the possibility of inscribing scientific activity within a closed horizon of total knowledge, they are close to Laruelle's ambition to separate science from philosophy and to eradicate the philosophical image of science in general. The Laruellian theorization of scientific objects as nonthetic, unidirectional reflections of the real, or as absolutely dispersed Unilateralities whose chaos cannot be integrated into any philosophical teleology or figure of totality, finds strong partial allies in the thought of these pluralist realists. Such a partial alliance is strengthened even further insofar as all of these three thinkers oppose scientism and in various ways reject the notion that science has a unique position of mastery or authority in relation to other regions or ways of knowing.

However, although Laruelle's realism-of-the-last-instance is arguably closer to open realism, promiscuous realism, and metaphysical nomological pluralism than it would be to other philosophy of science positions (e.g., strong realism, or conversely variants of constructivism or relativism), the alliance between them can only ever be partial, and the proximity can only ever be approximate. Indeed, from the strict Laruellian and nonphilosophical point of view, there can be no alliance at all because, as has been underlined from the outset, the positions of d'Espagnat, Dupré, and Cartwright remain philosophical. The thinking of the real as veiled is still a conceptual-representational construct. The ontological multiplicity of promiscuous realism is inscribed within the philosophical horizon of being, just as metaphysical nomological pluralism self-avowedly remains within the orbit of metaphysics. The question needs to be posed as to how the Laruellian nonepistemological sciences will be radically different from these realist pluralist positions. Laruelle would argue that it is only by way of his nonphilosophy or nonepistemology that any thought can be made strictly and rigorously adequate to the indivisible immanence of the real. By the same token, only nonepistemological thought can fully and adequately separate science from its capture by philosophy and its embeddedness within metaphysical horizons.

In this regard, what is most original and entirely unique about Laruelle's thinking is not just the radicality and rigor by which he seeks to eradicate philosophy from the natural sciences and to offer an image of what will in this chapter be called the open sciences. His originality lies also in the manner in which he seeks to extend the posture of immanence proper to science to all thought in general. This is the means by which he reformulates or reforms the field of knowledge per se, and in a manner that is far more thoroughgoing and transformational than anything thought or imagined by the other thinkers treated here. The new nonepistemological image of the sciences, so close to the images of science given by d'Espagnat, Dupré, and Cartwright, is therefore inseparable from a new image of thought in general, and this transforms all areas of knowledge. By axiomatically positing the real as an undetermined cause-of-the-last-instance of all phenomena, Laruelle renders everything contingent in relation to that last cause alone: the objects of scientific knowledge, nonscientific knowledge, and all possible thoughts or knowledge forms as such.[6] Laruelle expresses this as follows:

> "Phenomena" cease in their turn to be predetermined by Being or by some ontological finality and horizon, and since "experience" is delivered to its own *identity*-of-order, science is extended to any phenomenon that can now become the "object" of a science. Or, rather, that can *give place and indication to a new science*. The open multiplicity of the sciences has no other reason than this expansion, for which the Identity (of) the real is the cause and militates against every transcendent philosophical economy of "objective givens" and of fields of research. (1992, 92; 2016, 66, translation modified)

One could argue that there is a certain ordering of Laruelle's thinking insofar as it first begins with the posture of immanence proper to science and seeks an image of the natural sciences that will separate them from their philosophical capture and restore them fully to that posture of immanence to their structural essence as sciences. Having clarified the essence of science per se in relation to the sciences, nonphilosophical thinking can then extend itself as science, as a posture of immanence within thought, to all thought, phenomena, and knowledge in general. What this gives is "*two sciences in a dualitary correlation* through their respective object" (Laruelle 1992, 257; 2016, 208), with the first being the sciences stripped of philosophy and returned to their posture of immanence and the second being nonphilosophy itself—the general science of thought that takes philosophy as its object. Laruelle describes these two sciences in his own more idiomatic terms, arguing that there is "a science of the essence of science (or of 'empirical' sciences) and a science of philosophy. The first is a 'transcendental' science, an episteme-without-logos, a transcendental epistemic rather than an epistemology. The second is the science (as well as a critique and a new practice) of the philosophical Decision in the way it is included in the first science, which discovers its particular object within it: philosophy itself" (1992, 257; 2016, 208). The nonepistemological redescription of the sciences is in a position of antecedence or precedence with regard to nonphilosophy (understood as a science of philosophy) because it establishes the paradigm of the posture of immanence that nonphilosophy will generalize to all thinking, or, as Laruelle puts it, "A science (of) science must precede and include a science (of) philosophy" (1992, 258; 2016, 209).

Here the most difficult and challenging aspects of Laruelle's thinking come to the fore. If all thought and all phenomena are now to be related back to the real as their unilateral cause-in-the-last-instance, then does not everything—the objects of scientific knowledge and nonscientific knowledge; all phenomena, both real and imaginary; indeed, anything and everything that can appear—become a Unilaterality? If this is so, then how can one distinguish between scientific knowledge on the one hand (whose objectivity is caused by the real) and pure phantasms, ideological imaginings, aesthetic fictions, or fantasies, on the other? By Laruelle's account, all of these would be determined or caused in-the-last-instance by the immanent real, and we would appear to be without any means of distinguishing between them. Does this not lead thinking back to a relativism of the most corrosive and wayward kind, according to which there can be no distinctions of truth and value to be made between different kinds of knowledge or ways of thinking? This question becomes particularly pertinent when it becomes clear that Laruelle is constantly borrowing scientific concepts from the sciences proper and incorporating them within the framework and fabric of his general scientific or nonphilosophical treatment of philosophy. Most obviously there is use of the language of fractals in *Theory of Identities* and the massive use he makes of the language and conceptuality of QM in his thinking of generic science that has been developed from 2008 onward. In order to explore this relationship further, a close examination of Laruelle's recent conception of generic science is now needed.

Experience and Experimentation in Generic Science

In the most recent phase of Laruelle's thought, science becomes generic science, and nonphilosophy becomes nonstandard philosophy. As will become clear, the deep structure of his theoretical practice as an axiomatic thinking of radical immanence remains unchanged. To this extent, what follows will continue to refer to nonphilosophy as well as to nonstandard philosophy, and will treat the terms as more or less interchangeable. However, the style and language of Laruelle's late work can pose difficulties or challenges for even experienced readers of this already difficult discourse. In particular, the extended and systematic use of the language and conceptuality of quantum theory forms the dominant register of this phase and articulates the core of the major later work *Philosophie Non-standard* (2010c). Here the conceptual edifice of quantum theory is extracted from the realm of fundamental physics proper and becomes a kind of model for thought and its relation to immanence in general. The material or the flux of thought are treated as waves and particles, and quantum concepts are used to express the relation of thought to the real. Here Laruelle uses terms like "superposition." He also draws terms from algebra, such as "idempotency." All of this offers a means of formalizing or modeling nonstandard thought as an irreducible multiplicity placed into an irreducible and unilateral duality with the real outside of any totalizing horizon. It is in this context that the use of quantum concepts needs to be understood. His use of theological and Christological discourse and its collision (as a collection of thought particles) with quantum concepts in works such as *Christo-fiction* (2014) is another challenging example of this, as is the encounter with Badiou's ontology staged in *Anti-Badiou* (2011, 2013a). Arguably, this most recent phase opens with the publication of *Introduction aux sciences génériques* [Introduction to generic sciences] (Laruelle 2008). What follows will by no means attempt to offer an exhaustive account of this phase in its totality; nor will it give any full commentary of such large and imposing works as *Philosophie Non-standard.* It will, however, aim to offer a clear sense of how the thought of radical immanence is modeled in late Laruelle into a specific configuration of experimental practice, and how this orders the relations of the sciences, generic science, and nonstandard philosophy to each other.

In the fifth section of *Philosophie Non-standard,* Laruelle gives an enigmatic characterization of generic science: "[Generic science] associates, or makes an alliance between, a message that says itself to be mysterious (and which from its birth is mysterious), that is to say philosophy, and a key, quantum physics, which it is necessary to possess in order to decipher the message [. . .]. Their conjoined reception is necessary; no cryptographic analysis of the philosophical message is possible without the quantum key and vice-versa. This duality of [generic science], its secret and its sole concern, is irreducible" (2010c, 482). This characterization is enigmatic insofar as it obscurely or opaquely affirms that the relation between philosophy and quantum physics is an irreducible duality and also a secret. There is a double opacity here: that of the original philosophical mystery message to be deciphered, and that of the secret duality or conjoined reception of both philosophy and quantum physics. The relationship between philosophy and science is once again called into question by this obscure formulation of Laruelle's later thought. In order to shed some light on the obscurity and enigma of this characterization of generic science, the terms "experience" and "experimentation" within nonstandard philosophy will be explored. These two terms are central to any practice of or thinking about the natural sciences and the role they play in generic science, which illuminates the singular relation between philosophy and science in late Laruelle and the secret duality of generic science.

To begin this exploration, it is worth first turning toward the key: quantum physics. An example drawn from contemporary philosophical debates relating to quantum physics may be useful here. Unlike the positions discussed in the previous section. The philosophical perspective elaborated within this example has no proximity to Laruelle whatsoever. In *The Emergent Multiverse* (2012), Anglo-American philosopher of science David Wallace mounts an ambitious defense of the Everett interpretation of quantum theory. The Everett interpretation posits the real existence of parallel physical worlds, of an innumerable and unlimited multiplicity of universes (the "multiverse" of Wallace's title). Wallace's book is a major work of theoretical and mathematically orientated philosophy of science. Without going into too much technical detail, it is useful to outline a number of principal points relating to his argument and to his defense of the Everett interpretation of quantum theory.

The most important point is to note that Wallace's argument fuses three elements. First, he affirms a strong realist interpretation of the mathematical formalism of quantum theory. That is to say, he holds the equations of QM to be a formal description of reality such as it is, in itself, and independent of human beings (2012, 13, 38). He therefore poses a strong isomorphism between mathematics and the real (unlike d'Espagnat and all the other anti-Pythagorean thinkers discussed previously). Second, Wallace appeals to a philosophical naturalism inspired by Quine, aligning him broadly with the metaphysical naturalism of Ladyman and Ross (3). As will be recalled from the brief discussion of Quine in the introduction, what is most important in this context is the ontological and epistemological primacy that Quinian naturalism confers on scientific practice in the construction of the relation between philosophy and science: it is the epistemological sufficiency of science that simultaneously grounds philosophical-ontological knowledge and gives an epistemological foundation to our knowledge of reality in general. Third, Wallace proposes a specific resolution (he says dissolution) of the quantum measurement problem, arguing bluntly that "there is no quantum measurement problem" (1).

In order to clarify the logic and interest of this argument, a return to the famous thought experiment of Schrödinger's cat is in order. It will be recalled that the unfortunate cat in question is enclosed within a box with a radioactive uranium atom in one of its corners; in another is a detector constructed so that it will function for a minute only. During this time, there is a 50 percent chance that the uranium atom will decay and eject an electron. When the electron hits the detector, it will activate a hammer that will shatter a small bottle of poison, thus killing the cat. It is worth recalling here also that the Schrödinger equations describe the quantum system of the uranium atom as a wave function—that is to say, as a wave packet or electronic orbital where energy is concentrated. These equations, deterministic and linear, describe the uranium atom as a quantum entity to which the principle of superposition is applicable: its constituent particles exist in different states simultaneously and are therefore said to be superposed. This means that the atom exists simultaneously in two superposed states, intact and decayed, at the same time, and the famous fate of Schrödinger's cat thus necessarily imposes itself. Within the box, it is simultaneously both dead and alive. Yet this is so only up to the moment when, by way of a spy hole in the box, the cat is viewed by an external observer. QM affirms that observation at the macroscopic level of a microscopic system is also an intervention in that system. As soon as the quantum system is observed, there will be a reduction of the wave packet; that is, the quantum system will reduce entirely to that which has been measured. At this point, there is a decoherence of the superposition: the atom is no longer simultaneously intact and decayed, but is either one or the other.[7] From the act of observation onward, then, the cat is now either dead or alive, but not both.

From this famous case, elaborated from the mathematical formalism of QM and from the quantum measurement problem (reduction of the wave packet and decoherence of superposition at the point of observation), the Everettian interpretation and the multiverse thesis can be fairly straightforwardly explained. It is worth noting once again that, for Wallace, the Schrödinger equations describe a reality of the deterministic and unitary evolution of the wave function. Therefore, superposition would be a real existence of all the different states that are superposed within the quantum system. According to the Everett interpretation, observation or measurement of the system only gives us one of those states. Observation does entail participation in the quantum system and therefore decoherence, but it also entails branching (Wallace 2012, 88).[8] The other states nevertheless do still really exist, but they do so in other branches of physical reality that are inaccessible to us—that is to say, in other universes. Superposition therefore does not describe what is as yet indefinite or probabilistic but rather describes real multiplicity (Wallace 2012, 36–37). There is thus a universe in which the cat is alive and another branch, another universe, in which it is dead. This is what science and mathematics unequivocally tell us, argues Wallace, and, since naturalism tells us that it is science that should hold authoritative sway over knowledge of reality, the thesis of the multiverse must necessarily impose itself without equivocation. The measurement problem within quantum theory in effect does not exist since the observer of any given microscopic system measures only one branch of physical reality, only one universe: our own.

Standing back from this, one might make a number of observations from the Laruellian nonphilosophical perspective without necessarily evaluating Wallace's argument in relation to debates around, say, realism or instrumentalism in science, and without intervening in the disputes that the Everett interpretation incites among physicists and philosophers. In relation to the concepts of experience and experimentation that are under discussion here, it can be remarked that Wallace preserves them in their most classic form in relation to scientific practice. In so doing, he reinstalls a solid ground of certainty into the center of QM, where paradox and uncertainty have hitherto largely reigned. In the universe of classical physics, there is no measurement problem: the act of measuring does not significantly affect or intervene in that which is measured. According to a realist or objectivist account, the subject of scientific experience or observation is effaced in the experimental situation to give an objective representation of physical reality. In the quantum universe, however, the subject is not effaced but rather participates in the observed phenomenon, thus giving rise to the problems and paradoxes of QM.

For Wallace, the scientific subject remains entirely intact in the Everett interpretation, but at the price of being multiplied. The subject intervenes in the quantum system; there is decoherence and branching. Yet the observing subject keeps its position of objectivity according to the classic subject–object relation of the experimental situation. It is simply that that subject is observing only one of the branches of the system. The wave packet continues its unitary and deterministic evolution in many branches at the moment of observation and therefore many subjects will necessarily exist in a multiplicity of universes, all of whom can observe other branches without there being the least communication between these parallel worlds. The subject of experience in general remains intact in its transcendence over its own particular world, intact in its scientific rationality, and intact within the limits imposed on it by the philosophical naturalism Wallace endorses. As regards scientific experimentation, it keeps its privilege in relation to our objective knowledge of the world and of human-independent reality, but only at the price of not being able to observe other quantum branches or other universes. However, this apparent limitation is abolished in its turn thanks to the confidence Wallace places in the realism of the mathematical formalism of QM. The equations of mathematics here represent the real as it is beyond the limits of experimental practice. They unequivocally affirm the real existence of other worlds in quantum branches.

In effect, it is possible to discern here on the one hand a gesture of scientific and theoretical limitlessness: the innumerable multiplication of universes and the affirmation of an apparently limitless multiplicity of worlds. Yet on the other hand a gesture of decisive limitation and philosophical closure can also be discerned: the subject of experience remains within its limits, those imposed by scientific rationality and philosophical naturalism. Even though multiplied by a potentially infinite number of worlds, this subject is nevertheless each time held within the transcendence of a single world, within the horizon of its being, and within the closure of its system of signification. In the language of Laruellian nonphilosophy, it can be noted simply that the apparently limitless multiplicity that Wallace affirms nevertheless remains subordinated to the unitary figure of "the Whole and of philosophical form" (1992, 180). This figure unites mathematical-realist formalism, quantum theory, and the autosufficiency or autopositionality of philosophy in order to incorporate the multiplicity of the multiverse into a totalizing logic. In Wallace's argument, the relation of quantum theory and naturalist philosophy is a relation of mutual reinforcement or of reciprocal foundation of the kind described in the introduction: the naturalist position calls on science to regulate philosophical knowledge, and in turn the scientific theory that defends the Everett interpretation calls on naturalist philosophy to regulate its realist account of mathematics and of quantum theory. Within this circle of reciprocal foundation, philosophical Totality ceaselessly affirms its authority and power of metaphysical closure.

This example of Wallace's theory as elaborated in *The Emergent Multiverse* (2012) is not exactly the key that quantum theory offers in order to decipher the mystery of the philosophical message within Laruelle's generic science. However, it does offer a philosophical image, in the contrastive light of which the formalism of generic science and its use of quantum concepts can be clarified. In this way, Wallace's multiverse theory gives a useful point of reference in relation to which experience and experimentation in generic science can be understood. As has been clear throughout the discussion of Laruelle (and in the discussion of Unilateralities in particular), nonphilosophy consistently affirms real multiplicity, or the multiplicity of the real, outside of any philosophical teleology or possibility of closure.[9] It seeks to mount the most rigorous defense of multiplicity and to guard against or otherwise suspend the false multiplicities of philosophical Totality.

As was indicated in the discussion of Laruelle in chapter 1, nonphilosophy abandons the philosophical operation of determining, configuring or touching on the limits of thought. Nonphilosophy, insofar as it identifies and traces the invariant structure of the philosophical decision, shows how the delimitation of transcendence in relation to immanence (and the dyads or mixtures of the two that follow on from this) form the essence of philosophy's foundational gesture and its autosufficiency or autopositionality. With regard to nonphilosophy's "total critique" of philosophical limits, Rocco Gangle has remarked that such a critique "would both permeate and exceed the limits of its object without at the same time breaking those limits" (2013, 147). Nonphilosophy, or nonstandard philosophy, is neither a transgression nor an abolition of philosophical limits. Rather, it identifies the structure of those limits and the operations that determine them from a perspective exterior to those operations. It incorporates philosophical limits as materials within an entirely different practice. Contrasting the multitude of classical deterministic universes as theorized by Wallace and the universe of nonphilosophical science gives the following: on the one hand, multiverse theory offers a limitless multiplicity of worlds that remain nevertheless subordinated to Totality and to a gesture of limitation or closure that is philosophical through and through. On the other hand, the nonphilosophical scientific universe affirms a multiplicity or a real multitude (that of the real immanent One) from a radical delimitation without limit of all thought. It surpasses, while leaving them entirely intact, all the limits of philosophy.

The comparison and contrast between Wallace's multiverse and the nonphilosophical universe allows for a further specification of the role, or lack of a role, of limits in generic science and Laruelle's later nonstandard philosophy more generally. On this basis, the following hypothesis can be advanced: nonstandard philosophy has no limits except those nonlimiting limits that it gives itself in the construction of a specific experimental site, arena, or enclosure (*enceinte* in French) of thought and where thought is exercised as such. At the same time, it has a single or sole edge, border, or frontier (*bord* in French; "edge" is the best rendering of the term, but "border," understood as an edge as well as a frontier, sits better in English and will be also be used hereafter). This perhaps oblique formulation allows the singular usage of the terms *experience* and *experimentation* in nonstandard philosophy to be clarified and will also open a way for the obscure and secret duality of philosophy and quantum theory within generic science to be illuminated.

The problematic of experience in nonstandard philosophy needs to be thought in relation to its unique border, and that of experimentation needs to be thought in relation to the limits of its experimental arena or enclosure and the construction of this arena in relation to that border. As will become clear, the border in question is understood as a site that is an edge of the real but also as a unique frontier of thought with the immanent real. This border of thought with the real, as Laruelle repeatedly emphasizes, is unilateral or one way (recalling the unilateral logic of causality- or determination-in-the-last-instance in his earlier work). It is also universal in that it is a border for all thought, all transcendence of world, and the phenomena in any given world. This universality is therefore "generic" and is that which gives its name to Laruellian generic science. Yet if this border is everywhere within thought, while not being a limit as such, how can it be experienced, approached, traced, or circumscribed? It would seem that this is a border that cannot be made the object of an experience, nor approached, nor traced, situated, or localized in any way. In order to try and make this point clearer, it is worth signaling that in *Introduction aux sciences génériques*, Laruelle argues that the generic designates "a universality of unilateral intervention" and that this universal constitutes the unilateral border of all existing disciplines. It is "a border in relation to which it is impossible to say whether it is exterior or interior to the formations of any given knowledge since it is that which comes to disciplines and, through this coming, wrenches them from the ground of positivity that resulted in epistemology" (2008, 50). So the key operation of generic science is to place thought back into in a relation of contingency upon this border. It does so in such a way that this border becomes a force that comes from within the immanence of thought. Through the unilaterality of that force and of its power of intervention, this arrival tears epistemological sufficiency and positivity away from knowledge and all its disciplines, and transforms them without itself being transformed in its turn. Indeed, this sounds much like a reformulation of the nonepistemological redescription of the sciences of Laruelle's earlier work, although more directly and immediately generalized or universalized to all knowledge.

Laruelle is clear, though: the single border of generic science is not at all similar to the limits that philosophy gives itself. It is not a question here of determining the limits of possible experience or of a thinking of or at the limit (as it is in Nancy). It is neither an experience of a limit nor a limit-experience (as thought by Blanchot in *The Infinite Conversation* [1969, 1993]). On this point Laruelle could not be more explicit: the generic "abandons the touching of edges that philosophy engages in" (2008, 77). Everything is thus reduced to, or rendered contingent in relation to, this border or edge of thought, which is nowhere localized as either an interior or as an exterior, but which arrives from everywhere to extract or tear thought from any horizon of philosophical Totality. Yet if this coming from the border or edge of thought tears all the disciplines of knowledge away from their ground of positivity and epistemological sufficiency, then what becomes of experience and of the subject of experience?

So far, it should be clear that Laruelle's thinking of a universal or generic and unilateral border of thought is a repetition of his thinking of the real as autonomous and indivisible, as the DLI of everything that can be thought, everything that is, all experience and all knowledge. It could be said that the disciplines of knowledge and of all thought in general do not border on the radically immanent real but that they are the border or outer edge of the real. Here Laruelle's recent polemic against Alain Badiou, *Anti-Badiou*, is worth citing: "The radical is on the edge of the Logos but not as *its* edge; it is the Logos that is reduced to the state of the edge of the Real" (2011, 162–63; 2013a, 199–200). So the border, the sole and unique border, cannot be situated or localized. It is not an object of possible experience, and it cannot be approached or encountered because everything that *is* is reduced to the state of a border or edge of the real, determined in-the-last-instance by that real, and, in generic science, transformed by the unilateral force of the border or edge.

In this context, the figure of experience and of the subject of experience cannot remain untouched. They are also necessarily torn away or extracted from their epistemological sufficiency and transformed by the universality of this unilateral force. Because the real exercises its force of determination unilaterally and is not touched or affected in return, it cannot be said that the experience of a subject in its transcendence or its world-directed intentional consciousness is an experience *of* the real, as if the real could be the content of experience. Conscious experience is perhaps better described as experience *in* the real (as the edge or border of the real). As radical immanence, the real is, as always in Laruelle, anterior to or before all experience, as such as its cause or DLI. So the relation of the subject of experience to the real (that is, the relation going from the subject to the real) is in effect a nonrelation. Immanence, being radical, offers no dialectical mediation between experience and the real. Reduced to the state of being a border or edge, losing its transcendence and thereby falling into immanence, experience is added to the real as its border, but without affecting or touching it in return. Here Laruelle calls on a concept drawn from algebra, that of the idempotent. This is a mathematical operation that can be applied to a term many times over without modifying or changing the initial term. For example, 1 is an idempotent of multiplication: $(1 \times 1 \times 1 \times 1) = 1$. For Laruelle, the logic of the idempotent governs the indivisibility of the real and governs also the logic of addition of experience to the real. It is in this context also that his use of the term "superposition," drawn from quantum theory, needs to be understood. In *Christo-fiction*, Laruelle explains the specific sense he gives to the term "superposition" as follows: "There is superposition when immanence is through and through the same and traverses the instances of transcendence it brings rather than containing them; at the same time [these instances] do not change [immanence] by adding themselves to it" (2014, 102). Added to the real as its border or edge, experience is also traversed by the real in its force of unilateral intervention.

This thinking of experience—or it might more properly be said, this experience of thought—is an experience of nonexperience, and experience of something that never appears or manifests itself but that changes the status of all appearance and all manifestation. It might be objected, perhaps, that this remains nevertheless an experience, that of a conscious and thinking subject, the nonphilosopher, and that the real nevertheless remains the object of this experience even in its very indeterminacy because it is determined, as it were, in and by that indeterminacy. This objection misses its target, however, and does not fully take into account either the axiomatic formalism of this thought or the experimental and performative practice that it necessitates, for the whole nature or status of experience has been totally transformed. The importance of the reference to gnosis throughout Laruelle's later thinking needs to be underlined here. At the beginning of *Philosophie Non-standard*, he writes: "The generic is the contemporary and nonreligious form of Gnosticism" (Laruelle 2010c, 34). He also refers to *Christo-fiction* as a conjugation of science and theology "in a gnostic spirit" (2014, 11). And in *Introduction aux sciences génériques*, he highlights in the most explicit fashion the relation of gnosis to experience such as it is transformed by the force of the generic: "We understand this enfolded or additioned immanence as a gnosis and as a lived instance that avoids the one-multiple of the lived instances of consciousness which are still imbued with transcendence" (2008, 65). The experience of the nonexperience of the real is thus no longer a lived consciousness impregnated with transcendence (toward a world). It is a gnosis, and the subject that flows from that experience is not a subject preserved in its transcendence in relation to a world but rather an immanental subject. A change of vocabulary imposes itself here: Laruelle will talk of an "immanent or idempotent lived experience [*vécu immanent ou idempotent*]," but this lived experience is lived before all conscious transcendent experience of a world, prior to any horizon of being, and before thought. This is an experience of or within thought figured as gnosis, a knowledge of nonknowledge of a lived immanence that is never divided by transcendence.

John Mullarkey has justly described nonphilosophy as an unconditioned thought and as a "self-standing knowing or gnosis" (2012, 144). He is right also when he underlines the mereological relationship of all the diverse elements and gestures of nonphilosophical thought: they are parts of the real and not representations of the real. Each text of nonstandard philosophy is an experience of thought understood as a gnosis of this lived, idempotent immanence. As such, these texts are uniquely performative and experimental in their status. From the force of the unilateral intervention of its unique border, nonstandard philosophy transforms the figure of experience, makes itself a gnosis of lived or idempotent immanence, and uses the immanental subject that derives from this as an organon to formalize the axioms of immanence and construct its experimental arena or enclosure. The extent to which experience and experimentation in generic science are intimately linked should now be clear: the one necessitates and produces the other.

What exactly, then, are the limits that nonstandard philosophy gives itself in the construction of its experimental arena or site of thought? By definition, an arena or enclosure (*enceinte*) is something that circumscribes space or that traces a limit or boundary. Perhaps these limits are those imposed by the universal and unilateral intervention of the sole or unique border and by the irreducible duality of thought and the real that results from this. Such limits would therefore be imposed by the logic of unilaterality and the operation of idempotence. In this case, the limits of the experimental arena of nonstandard thought would be the axioms of immanence—axioms that form the generic matrix, which can be seen to function in diverse ways in the recent texts of nonstandard philosophy. Yet in the end, it is not at all certain that these axioms are limiting in any way in the traditional sense of the term since the experimental arena of generic science has as its goal a production and invention of thought that would be without limit.

The limits of the experimental arena of generic science are rather those of a theoretical space artificially constructed with the aim of transforming materials drawn from diverse and each time different domains of knowledge: philosophy, science, theology, photography, and so on, without limit.

The arena, site, or enclosure of thought is therefore a space constructed with the purpose of welcoming or receiving a specifically chosen set of materials and is constructed as a kind of matrix: the quantum space or collider of thought particles that is *Philosophie Non-standard*, the Christological space of *Christo-fiction* (2014) where the discourse of science and theology are conjugated, the arena of *Anti-Badiou* (2011, 2013a) where there is collision with and transformation of the Badouian ontology of the void. It can be concluded, therefore, that the limits that generic science gives itself are not limiting but rather productive and creative. So once again it can be affirmed that there is only one frontier, only one unique border here: the unilateral and universal border of the real that dualizes all thought. The only form of limitation in this context is that experience and thought here will never know anything of the immanence that is their cause; they will only ever be a gnosis of an idempotent "lived" (*vécu*). As Laruelle puts it in *Christo-fiction*, "The secret of gnosis must be conserved" (2014, 14). The only limitation on the experimental practice of generic science is that of this secret and of its preservation. This secret of gnosis, or more precisely this secret that is gnosis, is exactly the secret of an experience of dualized thought. It is the secret of the unilateral duality of thought and of the radically immanent real. This returns us to the beginning of this discussion: the duality of thought in Laruelle's generic science is both irreducible and necessarily a secret.

To conclude this exploration of experience and experimentation in generic science, a return to Wallace's multiverse and to the question of the epistemological status of quantum theory is in order. In relation to quantum theory, can it be said that nonstandard experimentation makes metaphorical use of its language and concepts? The answer is no, and Laruelle himself is explicit on this point. Metaphor, of course, being a form of analogy or substitution, always refers to the world and to its representations; it is a fundamentally representational function of language. The use Laruelle makes of quantum terms is nonrepresentational. It is rather formal, topological, or structural. It allows the presentation, production, or modeling of thought that has fallen into, or been thoroughly rendered contingent in relation to, immanence, a thought placed in a relation of unilateral duality with immanence. Treating the materials or the flux of thought as waves, as particles, and as being in a relation of superposition vis-à-vis the real is not to represent it as such but rather is a way of formalizing it in its irreducible duality and multiplicity without returning this multiplicity of flux to a thinking that unfolds within the philosophical horizon of Totality. So where Wallace's thesis of the multitude of the multiverse remains firmly held within the embrace of philosophical Totality, the multiplicity of the wave flux of thought remains, within nonstandard philosophy, outside of Totality, and outside of all its limits and gestures of limitation.

The question arises as to whether nonstandard philosophy can, in its turn, intervene in quantum science and touch pertinently on or transform Wallace's arguments. The answer is both yes and no. The use Laruelle makes of quantum language and concepts has, it should by now be clear, nothing to do with its use within the context of scientific and theoretical practice in the traditional sense of the term (i.e., within the context of fundamental physics). Nonstandard philosophy identifies the totalizing philosophical structure and pretensions of the multiverse theory and of the Everett interpretation of QM, but it also leaves that structure intact, untouched in its specificity as a scientific theory. Yet at the same time, of course, generic science, and the force of the generic (its unilateral and universal intervention), diminishes the autosufficiency and deep ground of epistemological positivity underpinning Wallace's thesis, making it, but in-the-last-instance only, equal with other forms of thought in relation to the real. The Everett interpretation has its opponents in the realms of quantum theory and of cosmology more generally. Above all, the instrumentalist interpretation is viewed by Wallace as its most important opponent since it does not have such a strong realist interpretation of the Schrödinger equations. According to this interpretation, the cat in the box is always either already alive or dead, but its actual state cannot be known except by the intervention of a measuring instrument. This means that quantum superposition is not a property of the real but rather a balance of mathematical probabilities only. Ontologically speaking, there are no quantum branches superposed as different universes. The multiverse would be, according to the instrumentalists, an idea of pure fantasy—and not a helpful one in Penrose's (2016) sense. Nonstandard philosophy does not seek to arbitrate between these two camps (realist and instrumentalist); nor does it seek to evaluate their respective merits scientifically. Since they are both philosophical, it would render them equal in-the-last-instance in relation to the radical immanence of the real. But nonstandard philosophy, being a realism-of-the-last-instance, could arguably and easily admit that, within the restricted sphere of scientific experimentation, there might come a day when both modeling and empirical results (i.e., scientific objects as Unilateralities caused by the real) would allow some kind of choice to be made between the two camps and for, say, the multiverse thesis to be disqualified or retained. This, at least, is a question that can be posed as such. It should be noted in this regard that nonstandard philosophy is nonepistemological but not antiepistemological despite its strong drive to reduce epistemological self-sufficiency within philosophy, thought, and knowledge more generally. So disputes and debates can resolve themselves within their restricted experimental spheres so that one model might be preferred over another on the basis of the Unilateralities or complex objects that have been modeled, even though the resulting implied metaphysical-philosophical positions will remain equal to any other position whatsoever, but in-the-last-instance only. This is the outcome of Laruelle's unique realism.

This is important to note because it must be constantly underlined that generic science and its way of configuring experience and experimentation are in no way a relativism in relation to the world or its reality. Rather, it is always and only ever a production and equalization in-the-last-instance in relation to the immanent real. In this context, it remains for Laruelle a defense of real multiplicity and of lived, idempotent immanence against the transcendence of the world and against philosophical Totality. It is the defense of a lived immanence that is prior or anterior to all world, all horizon of being, and before any kind of possible multiverse or cosmos. Wallace affirms that if the thesis of the multiverse is rejected, then it will be necessary to reformulate the mathematics of quantum theory—or indeed to invent an entirely new philosophy of science (2012, 110). The epistemological modesty or insufficiency that generic science imposes onto both philosophy and the sciences suggests that in the future, this entirely new philosophy may be possible, desirable, and indeed necessary. Yet from the Laruellian perspective, such a new philosophy would come only on the condition that it will be a nonstandard philosophy.

The Open Sciences

This discussion has moved from an outline of Laruelle's recasting of the relation of philosophy to science, through an extended exploration of his nonepistemological redescription of the sciences in *Theory of Identities*, to a comparative analysis of philosophy of science positions that are close to the image of the sciences resulting from that redescription. It has then moved on to explore the complex relation of Laruelle's later nonstandard philosophy and generic science to the language and concepts of quantum theory. The specific transformations undergone by the terms "experience" and "experimentation" in generic science show just how far Laruelle's nonphilosophical or nonstandard perspective differs from philosophy of science perspectives that still retain some claim to represent or touch on the real. These key terms of scientific theory and practice take on a scope and meaning that is entirely different from anything that could be envisioned outside of the nonphilosophical arena.

As for the natural sciences themselves, the picture of these that has emerged is both consistent and coherent and will be developed to its conclusion here under the concept of the open sciences. This is a picture that also remains in large part consistent with the images of science yielded in the thought of d'Espagnat, Dupré, and Cartwright discussed above. It is realist (albeit in-the-last-instance), pluralistic (albeit genuinely, rather than faux-philosophically), internally nonhierarchical, and infused in its essence with a principle of (non)epistemological insufficiency or modesty with regard to the real. It is this principle of insufficiency that detaches it from the philosophical image of science, from the logic of the philosophical decision and the horizon of Totality or totalization that that decision inscribes. The picture of the open sciences given here is diametrically opposed to Ladyman and Ross's naturalized metaphysics. Insofar as it is nonhierarchical, the PPC is rendered entirely redundant. There is no possibility whatsoever of a unity of the sciences reflecting a unity of the cosmos. The notion that mathematics (OSR) or information (ITSR) can give a privileged and positive access to the real is totally precluded by the axioms of radical immanence and the vision of the Real as an indivisible One. Yet being nonphilosophical, it does not set itself up as a rival to Ladyman and Ross or any other position. It has the claim to restore the sciences to their essence of science by transforming their relation to philosophy entirely.

The image of the open sciences presented here draws on the work and thought of Anne-Françoise Schmid, a French philosopher of science and close collaborator with Laruelle. Schmid is director of the research group Academos (in whose activities Laruelle has participated), which works on the theme of, as she notes in one of the group's publications, "new interactions between philosophies and sciences" (Schmid 2012, 9). The influence of Laruelle on this project, and on Schmid's thinking in general, is significant, but she is also an original and important figure in her own right, a distinguished philosopher of science, a highly accomplished scholar of Poincaré and Bertrand Russell, and a thinker whose knowledge runs deeply across both French and Anglo-American perspectives in science, philosophy of science, and philosophy more generally. Institutionally, Schmid has been situated in contexts that have brought her close to the practice and application of science rather than to the purely conceptual activities of philosophy departments. She has worked with teams of scientists in her capacity as visiting professor at the Institut National de Recherche Agronomique (Centre de Jouy-en-Josasand) and has held posts at the Institut National des Sciences Appliqués in Lyon and at the École des Mines de Paris.

In a recent collection of essays publishing some of the work of Academos and entitled *Épistémologie des frontières* [Epistemology of frontiers] (Schmid 2012), the proximity of Schmid to Laruelle is demonstrated clearly when she speaks of "a non-exclusive vision of the contemporary sciences," which gives "a new importance to the concept of identity [. . .]. This identity is not determined by the unity of a content, but by the relation of its content to the real." She then immediately adds: "This considerably modifies the classical positivism through which the 'specificity' of the sciences was sought" (2012, 290). The "content" of science (its object) is thus not a synthesis (of real thing and concept) but relates back only to the identity that is its cause. This exactly recalls the theory of identities elaborated earlier and is just one example of an instance when Schmid reprises the formulations of Laruellian nonepistemology. There are many other examples where it is clear that Schmid has internalized the theoretical critique of philosophical-conceptual representation that nonphilosophy offers (2012, 16–18). However she also develops these insights in a manner that Laruelle does not and that relates more directly to questions of scientific practice and to interdisciplinarity.

One of Schmid's key concerns in this context is the central role played by modeling in the sciences (*modélisation* in French). This is a concern that runs throughout her writing, and the theory of modeling she develops forms the core of her innovative account of interdisciplinary practice in the sciences more generally. So, for example, in *Épistémologie des frontières,* she notes that "modelling is one of the most widespread practices of the contemporary sciences" (2012, 21) before going on to define modeling itself as "an articulation of models—which, most of the time, are not drawn from the same discipline. Take 'a' model. It connects terms together. Let's suppose that each of these terms is itself a model: now we have an instance of modelling. Models are always declined in the plural. Modelling permits an articulation of heterogeneous models, of differing size and nature, amongst themselves" (26–27). The central point here is that the models that form any practice of scientific modeling do so in the plural, and that this plurality or heterogeneity is an intrinsic element of the modeling process. So in the modeling of any given scientific object or content (or Unilaterality as Laruelle would say), different models may be simultaneously at play: mathematical, chemical, biological, computational, and so on. One of the examples Schmid gives in this context is that of climate change, which has given rise to a huge proliferation of different models drawing on diverse disciplinary specialisms (298). The key theoretical point to retain here is that modeling, in its intrinsic plurality, is highly susceptible to interdisciplinary practice. It is also important to note that for Schmid, "a model is not a representation" (229). This may seem odd, but what is being indicated here is that a model is a construct involving a selection of variables and parameters and is therefore also and necessarily a simplified formalization rather than an exact replica. A model is only ever approximate or partial in relation to the real it seeks to model. In this respect, the definition of a model recalls the configuration spaces of Smolin's Newtonian paradigm, as well as the nomological machines of Cartwright. This means in more Laruellian terms that a model is never in an exact representational conceptualization of or synthesis with the real. It is a simulation or simulacrum.

The necessary plurality of models as they are deployed within the process of modeling, when taken together with their approximate or nonrepresentational status in relation to the real, means that the process itself is both dynamic and nonhierarchical in the assemblage of diverse elements that it constructs. In an earlier work *Que peut la philosophie des sciences?* [What can philosophy of the sciences do?] Schmid notes: "Modelling can be characterized as a construction produced in function of a determinate problematization of passages and exchanges between laws, models, measurements, experiment, simulations, the results of observation, or of any other pertinent element" (2001, 129). In this context, no one element (e.g., mathematics or empirical observation) will have any more privileged access to the real than any other since each is equally insufficient in relation to a real that precedes it while at the same time also causing or determining it in-the-last-instance. So the dynamic and nonhierarchical practice of modeling is not reducible to simple oppositions of theory and experimentation, or to linear procedures that move from observation and experimentation to inference, prediction, and further verification.

Modeling, for Schmid, therefore implies a pluralizing of scientific method: "With modelling science is considerably enlarged and takes on faces that are other than the hypothetico-deductive method or that of verification and refutation" (2012, 287). With this comes an opening and pluralizing of the structure and interrelationality of scientific knowledge more generally that Schmid characterizes in terms of "the opening of a scientific knowledge which functions with a multiplicity of ingredients each time rearticulated, in an instance of modelling, and not played out in advance around the opposition of theory/experiment" (286). In this context, all the different elements of scientific method and knowledge—for example, concepts, lawlike regularities, theories, experiments, results and data, hypotheses, and simulations—all also have their place on an equal footing in relation to the real (228). This in turn means that any phenomenon or object of scientific knowledge is susceptible or liable to modeling by way of a multiplicity of different models (297–98). Once again, the concept of a universal law of nature is suspended in favor of a more modest conception of lawlike regularity in which laws are interpreted only "in function of the scientific models in play" (Schmid 2001, 133, recalling Cartwright's nomological machines).

It should be clear how a strong conception of interdisciplinarity is necessarily entailed by this account of the central role played by modeling in science. What has effectively been argued here is that models and modeling, such as Schmid conceives of them, are entirely compatible with the Laruellian nonepistemological redescription of the sciences and that the nonphilosophical plurality of Laruellian science manifests itself in the plurality and nonhierarchical heterogeneity of modeling with regard to the method, content, and interrelationality of different scientific practices. Schmid herself underlines the extent to which the interdisciplinarity that is implied by this thinking of modeling is new and not simply reducible to a mere combination of separate disciplines. In this context, she directly invokes the language of Laruelle's generic science insofar as the nonhierarchical heterogeneity and disciplinary multiplicity of modeling are placed in relation to, or rendered contingent on, the universal and unilateral border or edge of the real—that is to say, the generic frontier: "The generic frontier is that which, whatever the combination of disciplines a problem supposes, can be affirmed in the articulation of any collection of heterogeneous forms of knowledge" (2012, 336).

Schmid's account of modeling and interdisciplinarity thus carries over the thinking or axioms of radical immanence and nonepistemology into its pluralized vision of the sciences and of scientific practice. This account incorporates elements from Laruelle's earlier thought as well as key aspects of his more recent generic science (e.g., the theory of identity, the anteriority unilateral causality of the real in relation to philosophy and knowledge, and the universality of the generic frontier). This means that the interdisciplinarity theorized here cannot be a simple combination of disciplines that remain intact with regard to their traditional forms of epistemological positivity of sufficiency. Rather, what is decisive is "the importance of having an immanent perspective on interdisciplinarity [. . .] one that can be opposed to its classical interpretation as a 'combination of disciplines' around a complex object" (Schmid and Mathieu 2014, 337). This notion of a "complex object" forms the hinge or pivot around which Schmid's thinking of modeling and interdisciplinarity can arguably be said to turn. A complex object, in Schmid's sense, would fit well into the Laruellian schema of Identity, real object, and object of knowledge understood as a Unilaterality. Recall that a Unilaterality represents nothing of the real but relates back to an unknown and unknowable real Identity as its cause. In this context, it could be argued, bridging Schmid with Laruelle, that a complex object is a Unilaterality approached via the plural and interdisciplinary procedures of modeling. For her part, Schmid defines the complex object in the following terms as "an object which is not entirely resolvable into a single discipline [. . .]. A complex object relates to collections of coordinates constituted in the procedures and approaches of diverse disciplines" (2001, 227). A complex object here is one that can only be approached or known via a multiplicity of models drawn from different disciplines. At the same time, a Laruellian Unilaterality, unhooked from the horizon of philosophical Totality and dispersed into a chaotic multiplicity, is uniquely suited to the plural, nonhierarchical, and nonrepresentational procedures of modeling as described by Schmid. Modeling and interdisciplinarity provide a vision of science and of scientific practice that is fully adequate to the status of Laruellian Unilateralities and to the nonepistemologial redescription of the sciences more generally.

It might be tempting, as this discussion reaches its point of conclusion, to align Schmid's account of modeling with Hawking and Mlodinow's model-dependent realism as elaborated in their 2010 work *The Grand Design*. After all, they, like Schmid, place models and modeling at the heart of scientific practice and appear, at first glance at least, to be similarly pluralistic in their account of the role models play in the sciences, and similarly modest also in their acknowledgment that a model can never be identical with the real or know it in anything other than an approximate way. As Hawking and Mlodinow put it: "There may be different ways in which one could model the same physical situation, with each employing different fundamental elements and concepts. If two such physical theories or models accurately predict the same events, one cannot be said to be more real than the other; rather, we are free to use whichever model is most convenient" (2010, 7). It might also be tempting to take Hawking and Mlodinow's assertion that "philosophy is dead" (5) as an (albeit entirely unknowing and unintentional) endorsement of the nonphilosophical project. At the same time, the authors of *The Grand Design* appear to be agnostic, for the time being at least, about the idea that successive models of reality across the history of science will "reach an end point, an ultimate theory of the universe, that will include all forces and predict every observation we can make" (7). This might suggest that they, like Laruelle and Schmid, wish to affirm the plurality of scientific knowledge and its models outside of any horizon of totality or metaphysical closure.

Yet these apparent similarities conceal fundamental differences, differences that highlight the uniqueness of the Laruellian nonepistemological redescription of the sciences and Schmid's thinking of modeling and interdisciplinarity that is so closely allied with it. For when Hawking and Mlodinow so dismissively state that philosophy is dead, they do so on the basis that it "has not kept up with modern developments in science, particularly physics" (2010, 5). This suggests a strong form of scientism and philosophical naturalism that, like the naturalized metaphysics of Ladyman and Ross, places fundamental physics in a supreme position of epistemological, and indeed ontological, authority in relation to both the real and to other areas of knowledge more generally. Hawking and Mlodinow may concede that models do not know reality absolutely, but the scientific way of knowing that they articulate has an autosufficiency and autopositionality that confers on it an epistemological and ontological privilege that would still be the domain of philosophy if only it had kept up with science. Philosophy may be dead, then, but this is so only insofar as science and fundamental physics have taken on the mantle of philosophical authority in relation to both the real and to all areas of knowledge more generally.

That this is so is borne out by the central role played within *The Grand Design* by the idea of an "ultimate theory of everything." Despite their professed agnosticism as to whether science will reach an "end point," the authors nevertheless do have a "candidate for the ultimate theory of everything," which, as they acknowledge, forms the basis of much of their discussion (Hawking and Mlodinow 2010, 8). The M-theory, as they call it, "has all the properties we think a final theory ought to have" and it has a unifying role in relation to the diverse subtheories that compose it (Hawking and Mlodinow use the metaphor of multiple maps of the world overlapping to give a unified world map): "The different theories in the M-theory family may look different, but they can all be regarded as aspects of the same underlying theory" (8). Underlying difference and multiplicity are sameness and unity of the kind sought after by Ladyman and Ross in *Every Thing Must Go* (2007). From this affirmation of the unity of the M-theory, and leaping rather abruptly from a scientific to a theological-metaphysical register, Hawking and Mlodinow go on to indicate that the M-theory will address the issue of creation; citing Douglas Adams (semi-ironically?), they note in relation to the "Ultimate Question of Life the Universe and Everything" that "we shall attempt to answer it in this book" (2010, 10). At this point, Hawking and Mlodinow's "model-dependent" realism emerges as a robust and full-throated naturalized metaphysics, a theory in which science has taken on the mantle of philosophy and hooked itself to the teleological, unifying, and totalizing horizons that were once philosophy's preserve. The vision *The Grand Design* gives of scientific endeavor and knowledge is aptly summed up by its title, and it is one that has a metaphysical-scientistic closure at its heart.

The Laruellian picture of nonepistemologically redescribed sciences, and the theory of modeling and interdisciplinarity that accompanies it, can, in in stark contrast with this vision, be characterized as radically open. The sciences and the processes of modeling that constitute scientific practice are not tied to the teleology of an "ultimate theory of everything" that will unite all theories and predict everything that can happen in the physical universe. Rather, they are related back solely to the identities of the real that are their cause or to the intervening force of the generic frontier of the real whose edge or border they are. The knowledge produced here is no less real, being determined in-the-last-instance by the real while leaving the real intact and undetermined. This will for many continue to be a strange and difficult idea, one whose unique form of realism is ultimately irreducible to more familiar philosophy of science perspectives. Yet placing scientific knowledge back into a relation of causation by the identities of the real, or unilateralizing it in relation to the generic frontier, situates that knowledge in an open field or terrain, one that is unbounded by any horizon or possibility of philosophical totalization or metaphysical closure. Science is thereby genuinely pluralized and stripped of the hierarchies imposed on it by philosophical teleology and metaphysical prejudice. The Laruellian nonepistemological sciences are internally nonhierarchical in the manner described with the help of Schmid's account of modeling and interdisciplinarity. Yet their openness is now one that affects all knowledge and all disciplines, be they sciences, social sciences, or arts and humanities. All knowledge is now situated in that open field or terrain and in relation to its sole and unique border: the generic frontier of the real. Within this open field or terrain, the only limits that the diverse disciplines of knowledge have are those they give themselves, independent of philosophy, in the construction of their disciplinary arena, site, or enclosure. Within the arena or enclosure of any given discipline, different ways of knowing the world of natural or artificial phenomena, of thought, and of all history and culture are produced. They all configure knowledge of the world, thought, and so on differently, and they are thus in no way the same or equal in relation to the world, the horizon of being, and the transcendence of that which appears. Yet they are all equal in-the-last-instance in relation to the real that is their immanent cause and of which they are all equally parts (i.e., they are all equally real). The way in which the constructed limits of different disciplines relate to each other is now also mobile and dynamic, nonhierarchical, and irreducibly plural. Yet to be absolutely clear, there is nothing at all here to deny the specificity of scientific knowledge and practice and its posture of immanence that produces a specific way of knowing the real. The extension of Laruellian nonphilosophical science, understood as a posture of immanence that can affect all thought and all knowledge, has the effect of placing everything back into immanence and thus situates everything within an open and unbounded field. Laruellian experimental thought, and the image of the open sciences it yields, therefore allow for a reformulation, transformation, and reinvention of the relation of all forms of knowledge to each other.

4 THINKING BODIES

BOTH NANCEAN SPECULATIVE NATURALISM and the Laruelle-inspired vision of the open sciences describe an image of thought and knowledge that is radically detached from any horizon of unity, totality, or completion. This separation of thought from the horizons of unity or totality has implications for one of the most fundamental questions and problems of both philosophy and science: that of the relation between the mind and the body. This relation can be construed in different ways as that of phenomenal consciousness to physical existence, of thought to matter, of subjectivity to objectivity, or of qualitative experience to quantitative knowledge. It is a question that has been central to the trajectory of American naturalism that was traced in the introduction and a key aspect of the eliminativist naturalized metaphysics of Ladyman and Ross discussed in chapter 2.

David Lewis in "What Experience Teaches" argues that any thoroughgoing or consistent materialism must necessarily deny the existence of "phenomenal information"—that is, the notion that phenomenal consciousness or first-person experience can bear or articulate information that is in any way distinct from the physical information that can be observed from the third-person scientific perspective and known according to the laws, structures, and relations determined by science. "If the Hypothesis [of phenomenal information] is false and Materialism is true," Lewis argues, "it may be that all the information there is about experience is physical information" (1999, 271). Indeed, he goes further in arguing that if there were such a distinct thing as phenomenal information, it would necessarily imply some kind of ontological dualism and therefore invalidate materialism: "There is no way to grant the Hypothesis of

Phenomenal Information and still uphold Materialism. Therefore I deny this hypothesis. I cannot refute it outright" (277).

It is perhaps not surprising, then, that in its ambition to seek out a total theory, philosophical naturalism has tended toward a physicalism that is resolutely eliminative with regard to phenomenality, consciousness, subjectivity, and qualitative experience in general. David Papineau has argued that physicalism, as a necessary consequence of naturalism, is "the thesis that all natural phenomena are, in a sense to be made precise, physical" (1993, 1). We should therefore "uphold the simple physicalist position that all mental states, including conscious states, are identical with or realized by physical states" (103). All mental states must in consequence reduce down to physical states and be knowable not just by science in general but by physics in particular. "Physics," Papineau holds, "is *complete*, in the sense that all physical events are determined by prior *physical* events according to physical laws. [. . .] We need never look beyond the realm of the physical in order to identify a set of antecedents which fixes the chance of any subsequent physical occurrence" (16). The horizon of unity and totality, which, it was argued in the introduction, is the fundamental medium of thought in which philosophical naturalism has traditionally tended to orient itself, is once again articulated in this understanding of the completion of physics. If everything is reducible to the physical, then it is physical science that provides the lower limit and completed knowledge to which everything can be reduced.[1] This completion, however, comes at the expense of recognizing the distinct first-person perspective of conscious, phenomenal experience, or of mental life.

The kind of materialism and physicalism promoted by Lewis and Papineau find its perhaps most thoroughgoing and uncompromising expression in the eliminative materialism of Paul Churchland and Patricia Churchland. Paul Churchland describes eliminative materialism as "the thesis that our common-sense conception of psychological phenomena constitutes a radically false theory, a theory so fundamentally defective that both the principles and the ontology of that theory will eventually be displaced, rather than smoothly reduced, by completed neuroscience" (1981, 67). Everything that we might normally associate with the first-person perspective of consciousness (intentions, propositional attitudes, beliefs, desires, and so on) is assimilated in this framework to the order of the "common-sense conception of psychology," or what eliminative materialism dubs folk psychology. Folk psychology, Churchland (1996) argues, deserves to be treated as an empirical theory that has been implicit to our everyday attitudes over the last two millennia. As an empirical theory, it must be viewed as would any other scientific theory and tested against scientific concepts and knowledge. The general thrust of eliminative materialism, then, is to view the perspective of folk psychology as analogous with scientific theories or hypotheses that have historically been found to be false (theories of light, of "vital spirit," alchemy, and so on) and therefore as ripe for being superseded by the scientific perspective, and by neuroscience in particular. In the words of Patricia Churchland, the aim of such a neuroscientifically oriented philosophy would ultimately be to invent "one unified grand theory of the mind-brain" (1986, ix).

There have been important challenges to this eliminativist perspective, of course. One of the most significant and interesting of these is the naturalist dualism proposed by David Chalmers in *The Conscious Mind: In Search of a Fundamental Theory* (1996). In this seminal work, Chalmers accepts that "there is good reason to believe that *almost* everything in the world can be reductively explained," but, he argues, "consciousness may be an exception" (xv). Chalmers argues that consciousness is in some very real way irreducible because "a mental state is conscious if it has a qualitative feel—an associated quality of experience" (4). From this perspective, the Churchlands are making a fundamental error in treating folk psychology is if it were just another empirical theory that will be superseded in the way that alchemy and theories of light or life have been superseded within the history of science. Whatever folk psychology may or may not be by way of a set of concepts or assumptions relating to the first-person perspective of consciousness, what is at stake here is consciousness itself as the site or locus of experience that is lived as such and that is therefore ineliminable and indubitable. As Chalmers puts it in what can be read as a direct riposte to the work of the Churchlands: "Eliminativism about conscious experience is an unreasonable position *only* because of our own acquaintance with it. If it were not for this direct knowledge, consciousness would go the way of the vital spirit [. . .] there is an *epistemic asymmetry* in our knowledge of consciousness that is not present in our knowledge of other phenomena" (102). In this context, the question therefore is not that of the future elimination from theory of consciousness or qualitative first-person experience understood as a task on a par with the elimination of alchemy or vitalism within the history of science. Rather, the question becomes that of how and why natural and biological systems produce something like consciousness.

This is a question that is a central preoccupation of and fundamental concern for Malabou and Stiegler. In both cases, the manner in which we understand the relation of thought to materiality and embodiment has profoundly social, political, and cultural implications. In both cases, there is an imperative embedded in their philosophical projects to think through the relation of the phenomenal to the physical, of thought to matter, of qualitative experience to quantitative knowledge in ways that avoid reductivism and eliminativism while still embracing materialism. In both Malabou and Stiegler, the alternative put forward by Lewis of either embracing materialism and rejecting the "Hypothesis of Phenomenal Information" or granting that hypothesis and therefore rejecting materialism is itself rejected. This rejection can be discerned in the way that each engages with scientific discourses and does so to produce a naturalism that is without an eliminativist or reductivist orientation but that nevertheless rejects any kind of dualism, naturalist or otherwise.

Plasticity and Epigenesis

It is in this context that Malabou's book, *What Should We Do with Our Brain?* (2004b, 2008), explicitly brings together the apparently incompatible perspectives of her distinctive postphenomenological and postdeconstructive thinking of plasticity with that of neuroscience. Indeed, the wider challenge posed to philosophy by recent discoveries within neuroscience is taken up in Malabou's work in a way that seeks to synthesize these opposing modes of thought. In so doing, it opens the way to thinking beyond the oppositions implied in the physicalism and eliminativism of Lewis, Papineau, and the Churchlands on the one hand and the naturalistic dualism of Chalmers on the other. In this light, Malabou's distinctive, what will be called epigenetic, naturalism opens up radically novel perspectives within recent and contemporary neurophilosophical debate.

What Should We Do with Our Brain?

Given the apparently stark incompatibility between the postdeconstructive and the neuroscientific perspectives, Malabou's thinking about cerebral plasticity in *What Should We Do with Our Brain?* (2004b, 2008) raises some difficult and pointed questions concerning the ways in which any postphenomenological philosophy might engage with the empirical findings of neuroscience. These questions relate to the specific ways in which Malabou affirms her distinctive form of philosophical materialism. She insists throughout her work that mind, consciousness, and mental life exist in a continuity with the operations of the brain, with its neural networks and systems and their synaptic connections, and insists also that this has become a certainty that recent research in neuroscience has rendered incontrovertible. In this she is, or clearly wants to be, a physicalist and materialist, albeit in a different mold from Lewis and Papineau. In fact, she follows neuroscientists such as Antonio Damasio and Joseph Ledoux in positing the idea of a neuronal personality, a neuronal self, or protoself. In this regard, she sets herself in direct opposition to Patricia Churchland, say, who is unequivocally eliminative with regard to the concept of self (1986, 407). At the same time, Malabou aims to think brain plasticity and the relation of the neuronal to the mental from the perspective of her earlier elaboration of the concept of plasticity, which she began in her work on Hegel and then developed further in her engagement with Heidegger. The crux of the difficulty or problem here relates to the way in which empirical scientific insights relating to brain plasticity are engaged within a naturalist and materialist perspective, and are then synthesized with the antiempiricism of the posttranscendentalist, phenomenological, and postphenomenological tradition. At the end of her book, Malabou addresses this question indirectly by suggesting that when it comes to thinking the relation between the mind and the brain, or the mental and the neuronal, the debate between reductionist and antireductionist approaches is a futile one, and one that philosophy must now seek to avoid. She continues: "One pertinent way of envisioning the 'mind–body' problem consists in taking into account the dialectical tension that at once binds and opposes naturalness and intentionality, and in taking an interest in them as inhabiting the living core of a complex reality. Plasticity, rethought philosophically, could be the name of this *entre-deux*" (2004b, 164; 2008, 82). Malabou explicitly proposes that the brain, taken as an object of empirical natural science on the one hand and as intentional consciousness as studied by phenomenology on the other, needs to be understood in a complex dialectical relation of terms. The neuronal and the mental or the cerebral and the intentional exist in a relation of continuity to each other, but that relation is far from simple, and the passage of one to the other needs to be philosophically elaborated or interpreted. She proposes that the concept of plasticity, both neuronally and philosophically reconfigured, will allow us to think this dialectical relation. The phenomenologist who would claim that empirical observation can tell us nothing whatsoever of intentional conscious states, and the neuroscientist who would affirm an absolute transparency between what is observable empirically in brain states and the experience of consciousness are both wrong. We need to find a way of thinking the two together.

It is this perceived insufficiency of purely reductionist or antireductionist approaches that therefore clearly motivates the meeting of naturalist-empirical and transcendental-phenomenological traditions in Malabou's work. In order to discern some of the difficulties, problems, or limitations of her position, a more detailed elaboration of her philosophical interpretation of neuroplasticity is now needed.

As has been indicated, Malabou follows Damasio and Ledoux in positing the existence of a neuronal personality or self. In particular she follows Ledoux in arguing that the structuring of our neural networks and synaptic connections provide the map of what we are in terms of our identity or self, and that the plasticity of these networks or connections means that the self is always formed in relation to its capacities of learning and of memory, and in its exposure to experience. On this she is explicit; with neural plasticity, "the entire identity of the individual is in play: her past, her surroundings, her encounters, her activities; in a word, the ability that our brain—that every brain—has to adapt itself, to include modifications, to receive shocks, and to create anew on the basis of this very reception" (Malabou 2004b, 20; 2008, 7). This position leads to a thoroughgoing questioning of essentialist understandings of identity or behavior that would appeal exclusively to evolutionary adaptations of brain structure and to notions of hardwiring. Brain plasticity, Malabou argues, implies not only that neural networks constitutive of personal identity are formed and created in and through experience, but that they also can never exclusively obey any given, received, or already constituted form that would circumscribe or determine identity. Here Malabou discerns the basis for a new thinking of freedom, and also for a political and ideological critique of the various discourses that surround and are connected to modern contemporary neuroscience. Brain plasticity means that our identity or self can always, throughout life, be a question of modeling or sculpting, of altering or modifying what we are in and through experience without recourse to any preconceived model, essence, or predetermined adaptive structure.[2]

What is at stake here is the way in which Malabou's account of neuroplasticity is cast in terms of the dialectic between the neuronal and the mental or the cerebral and the intentional. In order to assume the freedom that brain plasticity makes possible, we need to come to what Malabou calls a "consciousness of the brain" (that is, of its plasticity) and to assume that consciousness as a project. As Malabou puts it: "The consciousness we want to raise on the subject of plasticity has to do with its power to naturalize consciousness and meaning" (2004b, 24; 2008, 9). Assuming the thought of plasticity means, in the first instance, unequivocally accepting the ontological continuity of the cerebral and consciousness, of the neuronal and the mental. Yet for Malabou, such an assumption is by no means straightforward. Although she clearly accepts that "the transition from the neuronal to the mental is confirmed by the fact that it is impossible to distinguish between the two domains," she also adds, correctly: "Despite the apparent assurance and certitude which governs the discourse of the 'adherence' of the mental to the neuronal, the process of the 'translation' of contents from one realm to the other remains obscure" (2004b, 127; 2008, 62). This is indeed the key problem, one that was foregrounded in the discussion at the beginning of this chapter insofar as naturalist thought was shown to have responded to such obscurity either by seeking to eliminate consciousness as a scientific category altogether (the Churchlands) or by persisting with some kind of dualism (Chalmers).

This in fact thoroughly obscure process of translation from the realm of neuroscientifically determinable/measurable cerebral activity to that of first-person conscious or mental experience is not just a question or problem for philosophy and theory. It is also a problem, arguably, for empirical research programs and techniques within neuroscience. We know, for instance, that functional magnetic resonance imaging (fMRI) permits exact measurements of blood flow in the brain, and therefore of brain activity in various neural systems. We know that this activity can then be correlated to observable behaviors in specific situations. Yet it is also the case that what is given under these experimental circumstances is just that: correlations—and correlations that need to be supported with intuitive ideas, theoretical hypotheses, inferences, and interpretations, few of which will have a strictly empirical or empirically grounded status.

This issue has been the subject of some controversy within the field. In 2009 a controversial meta-analysis of existing scientific research was published entitled "Puzzlingly High Correlations in fMRI Studies of Emotion, Personality, and Social Cognition" (Pashler et al. 2009).[3] On the basis of a study of fifty-five papers that reported research carried out by correlating diverse emotional, social-cognitive, and personality-based behaviors with neural imaging data, this article uncovered serious anomalies in the techniques and procedures that were used across a wide range of experiments designed to detect whether such behaviors or experiences correlated strongly with observable brain states. The initial premise of the investigation was that the scientific articles surveyed, many of them published in high-impact journals and widely reported by a sympathetic media, were reporting implausibly high correlations as the basis for their inferences about the neural basis of social behavior and cognition. It was discovered during the course of the research that, in hidden ways not reported in the scientific papers themselves, many of the papers based their correlations on nonblinded (or nonindependent) selection of the neuroimaging data, thus yielding correlations that, although interpreted as being highly significant, were in fact likely meaningless. Given the nature of social neuroscience research and its reliance on neuroimaging (and therefore on supposedly solid and scientifically grounded interpretative findings), the implications of the hidden data selection methods were clearly serious. For instance, one study that correlated self-reported jealously in men and women with neural images found that activity in the insula showed a high correlation with reported infidelity-related jealousy in men but a low correlation in women; further investigation found that the data block from the fMRI had been selected differently in each case, nonindependently in the case of men and independently in the case of women (Pashler et al. 2009, 282).

With its 2009 publication date, the "Puzzlingly High Correlations" paper appeared in the relatively early days of fMRI imaging and took pains to indicate how the methodological defects of the studies surveyed could be corrected with more rigorous data selection techniques. Things have no doubt greatly improved in the field since then. Yet what the paper arguably also pointed toward was a problem, or at least a question, raised by research based on the correlation of brain imaging to experiential conscious states in general. Correlations cannot be reached independently of intuitive ideas, theoretical hypotheses, or inferences that are not always strictly scientific in nature (e.g., that men are less emotionally tolerant of infidelity than women, and that this difference may have a biological basis). That the research discussed in this paper should have presupposed possible correlations in their nonblinded selection of neural imaging data was, it would seem to the layperson, a clear breach of scientific method and protocol. Yet although high and reliably measured correlations are deemed to be both significant and more secure in this kind of research, the question that might be posed from the perspective of philosophy is whether the reliance on supporting frameworks, hypotheses, and so on might not always and per se be a problem for the interpretation of overall results (that is, whether there will not always be an undue conflation of correlation and causation).

A good example of this can arguably be found in the results of some experimental research published in 2013 in the scientific journal *Neuron* entitled "In the Mind of the Market: Theory of Mind Biases Value Computation during Financial Bubbles" (De Martino et al. 2013). Here the brain activity of financial market actors was monitored and the results were then interpreted on the basis of specific "theory of mind" assumptions, with conclusions arrived at concerning the behavior of market actors, the development of financial bubbles, and, most significantly perhaps, the evolutionary adaptions of brain structure that motivate or determine such behavior. Even if one assumes that the neural imaging data in this study were selected on the basis of a far more sound methodology than those studies surveyed in 2009 by Pashler et al., questions arise here. They arise in relation to the way the "Mind of the Market" experiments seamlessly married neural imaging data with assumptions drawn from one specific branch of cognitive science (theory of mind) and with assumptions drawn from evolutionary psychology as well as with, perhaps, unreflective assumptions relating to the kinds of agency and rationality that underpin market behavior or activity in contemporary capitalism (and this only five years after the global financial crash of 2008). When understood within the wider theoretical or philosophical framework of the desire to naturalize and reduce complex social behavior to its physical or biological base, such correlation-based research must surely be questioned with regard to its overall social, political, and ideological role or significance, and in particular with regard to the role played by the interpretation of correlated neural imaging and reported experiential states or behavior.

This brief foray into the wider research field of brain imaging and social neuroscience is useful because it brings into focus both the wider stakes and the core philosophical problem of Malabou's interrogation of the relation between the neuronal and the mental. It demonstrates why she is right to insist that we cannot assume a simple transparency between observable brain activity and the intentional states of consciousness to which they can be correlated. To do so is to refuse to acknowledge that a large measure of interpretation or hermeneutic activity that is not reducible to what is empirically observable is also engaged in such experimental situations. It is to risk falling into the worst kind of unacknowledged ideological discourse (which arguably the jealousy-related study reported by Pashler et al. in 2009 does unequivocally, and which the "Mind of the

Market" study from 2013 risks doing). This necessity of interpretation or hermeneutic activity leads Malabou to argue that the continuity between the neuronal and the mental is itself both neuronal and mental, biological and cultural; it is an "object of observation and an interpretative postulate," and as such "in essence a theoretical mixture, a simultaneously experimental and hermeneutic instance" (2004b, 128; 2008, 62). Malabou's analysis viewed in the light of the experimental studies examined above offers a strong account of what remains obscure in any attempt to think the passage from the neuronal to the mental and motivates her attempt to think this passage in terms of a complex dialectic.

Yet it is not entirely clear in *What Should We Do with Our Brain?* whether her rather provisional and sketchy account of this dialectical relation entirely clarifies this obscurity or resolves all the difficult issues her arguments raise. If we return to the account she gives of the neuronal, then the manner in which she conceives this dialectic becomes clearer. The neuronal self is plastic; it is held between the search for an equilibrium or constancy of form that gives the self a distinct identity but that also is exposed to possibilities of transformation and remodeling, to the alterity of a world of experience that can reshape it. Recalling Antonio Damasio's rejection of the idea that plasticity might entail a "wholesale modifiability" of neural structures, Malabou argues that "the transition from the neuronal to the mental supposes negation and resistance. There is no simple and limpid continuity from one to the other but rather the transformation of one to the other on the basis of their mutual conflict" (2004b, 147; 2008, 72). It is not a question here of conscious intentionality simply willing alterations in the connections of neural networks but rather of the way in which neural formations, despite their propensity to remain stable and to maintain their form, can nevertheless be modified by exposure to experience. This is most obvious in simple examples of learning. Learning to play the violin or learning a foreign language require prolonged and repetitious activity before becoming automatic and un-self-reflexive, brain-wired behavior. Similarly, the activity of certain neural networks might determine the contents of intentional consciousness in ways we would experience as negative or beyond our volition or control, as is the case in addiction, depression, or compulsive behavior. This is why neuroplasticity in Malabou's account is dialectical: the passage from the neuronal to the mental is one of conflict and negation. Each instance resists but can also be negated and therefore transformed by the other in the constant conflictual interplay between the neuronal and the mental, the cerebral and the intentional, between the empirically observable material thing that is the brain and the nonreifiable life of consciousness.

However, in describing a dialectical relation of conflict between two separate instances, it is questionable whether Malabou has really adequately, or in fact, described the passage from the one to the other, or even, really, the ontological continuity of the one with the other. For all her assertions of naturalism and materialism, the positing of the mental and the neuronal as separate instances that relate to each other in terms of resistance and conflict appears to presuppose some kind of separation between them, to place them into two distinct realms and therefore imply some kind of ontological dualism. Indeed, she explicitly concedes this point by suggesting that there exists some kind of ontological breach or gap separating the two: "Only an ontological explosion could permit the transition from one order to the other, from one mode of organization to the other" (2004b, 147; 2008, 72). The mental and the neuronal here are to different schema, two different "dispositifs" that can be dialectically transformed one into the other, but only in a "transformation which necessitates a rupture, the violence of a gap or opening which interrupts all continuity" (2004b, 149; 2008, 73). Malabou's complex relation of continuity between the neuronal and the mental as dialectical and conflictual is thus also a relation of violent discontinuity, of ontological rupture, and of the opening up of a breach or gap between what is empirically observable in the brain and what is experienced by intentional consciousness.

This is consistent with her argument that any approach to this question will always be a theoretical mixture between the empirically observable and the hermeneutic, the biological or the cultural, since both instances are always, one might say, always already engaged when one talks about brain plasticity, the neuronal self, or cerebral activity. Yet rather than shed light on how the mental might have its genesis in the neuronal or arise out from it, it appears to treat the two as ontologically distinct realms, always in relation to but nevertheless separate from each other. The question necessarily arises, therefore, as to whether Malabou, despite her insistence on naturalism, materialism, and ontological continuity, is not nevertheless reintroducing some kind of mind–body gap or ontological difference into her account, and whether she is not smuggling into her discourse an assumption of the transcendence of consciousness over brain activity at the precise point at which she seeks to disavow such transcendence.

The metaphorical language of ontological explosion, gap, or rupture that Malabou uses in the context is perhaps ultimately almost as obscure as the relation of passage from one instance to another which she is attempting to elucidate. This criticism needs to be developed in further detail. Yet when Malabou argues that "it is necessary to suppose, at the very core of the undeniable complicity that ties the cerebral to the psychical and the mental, a series of leaps or gaps" (2004b, 153; 2008, 75), she raises questions that her short, often polemical work and her invocation of dialectical transformation cannot answer.

Malabou notes that the idea of a neuronal self, such as it is advanced by Antonio Damasio, is inseparable from his hypotheses relating to the autoregulation of brain systems by way of modes of autorepresentation or self-representation. For Damasio, the argument that the self arises from a perpetually re-created neuro-biological state is central to the way in which he attempts to think the relation of mind to body. In *Descartes Error: Emotion, Reason and the Human Brain (1996)*, for instance, he argues that the brain and the body need to be viewed as an interrelated system that not only interacts by way of stimulus and response with its environment but, crucially and also, creates internal images or representations of those interactions. In a number of complex ways, the brain's neural activity is representational and symbolic, and is so as a function of its fundamental adaptation to its environment. Malabou takes this insight up in the following terms: "There is a self-representation of the brain, an autorepresentation of cerebral structure that coincides with the autorepresentation of the organism. This power of representation, internal to neuronal activity, constitutes the prototypical form of symbolic activity. It is as if the very connectivity of connections—their structure of referral, that is to say their semiotic nature in general—represents itself, or 'maps' itself, the representational activity being that which, precisely, blurs the frontiers between the brain and the psyche" (2004b, 122–23; 2008, 59–60). The origins of self and self-consciousness, what Damasio calls the proto-self, could perhaps therefore be located in this fundamental operation of cerebral autoregulation and autorepresentation. What might be decisive is that the activity described here is symbolic or semiotic; that is, it is a dimension of sense and meaning that is generated from or out of the brain–body's interactions with its environment in relations of meaningful inscription and referral. This would lead to the conclusion that Malabou posits but does not develop much, namely, that "meaning, or symbolic activity in general, depends strictly on neuronal connectivity" (2004b, 127; 2008, 62).

It is at this point that Malabou's postdeconstructive thinking opens up, yet leaves unexplored, a passage toward biosemiotics and the notion that biological agency is predicated on the naturalistic establishment of sign relations. The tradition of biosemiotic thought (briefly evoked in relation to Nancy and Canguilhem in chapter 2) can be traced back to the late nineteenth and early twentieth centuries and work by figures such as Charles Sanders Peirce and Jakob von Uexküll (Favereau 2010). It also inflects the work of Damasio and his notion of a neuronal self that lies at the center of Malabou's reflections throughout *What Should We Do with Our Brain?* The biosemiological insight that "the minds of human beings are themselves the product of a *de novo* use of absolutely natural and biological sign relations" (Favereau 2008, 11) offers a means of connecting the dimension of intentional consciousness, meaning, and experience to the neurobiological processes of the brain without reverting to ontological dualism (naturalist or otherwise) and without engaging in reductivism or eliminativism with regard to the former.

So rather than appeal to a modified Hegelian dialectic in order to think the passage from the neuronal to the mental or from the cerebral to the intentional, it might be more fruitful to develop the insights relating to the neuronal self and the symbolic autorepresentation of brain–body activity in terms that further connect the postphenomenological thinking of sense elaborated via Nancy and Canguilhem in chapter 2 with biosemiotic approaches. It will be recalled in relation to this earlier discussion that the notion of sense, understood as the relational being-in-the-environment of living organisms, was aligned with Nancy's thinking of phenomenality or world disclosure on the one hand and with the notion of physical information on the other, thus placing qualitative subjective experience and quantitative objective knowledge on the same ontological footing. When in turn Malabou affirms neuroplasticity as a "a power to configure the world" (2004b, 82; 2008, 39), it is difficult not to pass immediately onto Nancy's thinking of sense as phenomenal world disclosure and as the sensible-intelligible interactions and contact of material bodies with their surrounding environment.

Yet if Malabou's *What Should We Do with Our Brain?* (2004b, 2008) does stand as an opening of a way toward the synthesis of postphenomenological and biosemiotic perspectives (thus bridging opposing transcendentalist and empiriconaturalist trajectories of thought and knowledge), the book has arguably been shown to raise more questions than it answers, and it does not develop further some of its key insights and their implications. Most important in this regard is the question of status of the biological in relation to the transcendental (or, if you like, the postdeconstructive quasi-transcendental) and of how this relation is to be thought. Although the opening onto the biosemiotic perspective points the way to some kind of resolution of this question, it is left problematically open in *What Should We Do with Our Brain?* In Malabou's later work, *Before Tomorrow: Epigenesis and Rationality* (2014, 2016), the relation of the biological to the transcendental it is dealt with head-on and forms its central philosophical preoccupation.

The Place of the Void

As was shown in chapter 1, the concept of plasticity both offers an overall image of philosophy that informs Malabou's practice of thinking and constitutes a guiding thematic at each and every stage of her development as a thinker. The discussion in chapter 1 of plasticity as a general schema of thought in Malabou also concluded that such a schema was not so much a transcendental or grounding principle for philosophy as it was an experience of the groundlessness or absence of foundation within thought (just as freedom was for Nancy in *The Experience of Freedom* [1988, 1993b] and the default of the origin was for Stiegler in the first volume of *Technics and Time* [1998]). It is this experience of an absence of foundation, of an ontological groundlessness or void, that, it will be argued, offers the key to understanding the relation of the empirical to the transcendental in Malabou's thought as a whole. Throughout Malabou's philosophical work, plasticity bears witness and testifies to the void, the groundlessness, the abyss of being, and the void is always plasticity's most originary moment. Clarifying this point allows for the resolution of the unanswered questions of *What Should We Do with Our Brain?* (2004b, 2008) and yields the key to understanding her distinct form of materialism and naturalism.

It will be remembered that Malabou consistently refuses the notion of a gap or break between the regime of subjectivity, consciousness, or thought on the one hand and material, biological, and embodied existence on the other. In her book on Kant, rationality, and epigenesis, *Before Tomorrow*, Malabou asks: "Why does philosophy continue to ignore recent neurobiological discoveries that suggest a profoundly transformed view of brain development and that now make it difficult, if not unacceptable, to maintain the existence of an impassable abyss between the logical and biological origin of thinking?" (2014, 2; 2016, 1). This question reiterates Malabou's long-standing conviction that there is an ontological continuity between the neurobiological or neuronal dimension of thought and the experience of thought itself as consciousness or as the lived subjectivity of mental life as such. It reiterates the certainty, so clearly articulated in *What Should We Do with Our Brain?*, "that there exists a perfect continuity between the neuronal and the mental" (2004b, 114; 2008, 55). Such a question, formulated in this way, would appear unequivocally to indicate that there is no place at all for any conception of the void in Malabou's materialism. The absence of "an impassable abyss" between the logical and biological origin of thinking might well lead one to question where, after all, there can be a space or place for any kind of abyssal opening or vertiginous absence of ground within being when there is nothing but a seamless continuity between thought and matter.

The philosophical problem here, the relation of the empirical to the transcendental within Malabou's marriage of postphenomenological thinking and neuroscience, is the fact that, as she herself puts it, "the neurobiological viewpoint simply erases the transcendental" (2014, 262; 2016, 152). The truth of this is no doubt borne out in the reductivism and eliminativism of Lewis, Papineau, and the Churchlands: seemingly the neurobiological or neuroscientific perspectives have no need at all to pose the transcendental question of the formation of reason and rationality; nor, seemingly, do they feel the need to account for the phenomenological problem of the transcendence of intentional consciousness toward a meaningful world. Malabou develops a series of arguments against reductivism in *Before Tomorrow* and argues also for the philosophical inadequacy of any attempt to dispense with the transcendental moment within thought, either from the neurobiological or from other contemporary theoretical perspectives (2014, 2016, chap. 12). What is of central concern here is the place of the void in Malabou's thinking, of the ontological groundlessness that is plasticity's most originary moment. The void has its place insofar as it makes possible the intrication of biological life and the transcendental—the thinking together of the neuronal and the mental without any recourse to a breach, separation, or split between mind and body.

In fact, notions of vacuity, negation, and void have been at the center of Malabou's thinking of plasticity from her earliest work on Hegel. In *The Future of Hegel* (1996, 2005a), Hegel's own use of the German term *plastizität* is mapped onto the dialectical operations of subjectivity and its temporal experience, and, Malabou shows, the moments of vacuity, dissolution, and annihilation are of central and paramount importance for the plastic character of the dialectical process. This is brought out most evidently in her reading of Hegel's interpretation of divine kenosis, that process described in Paul's Epistle to the Philippians, as Christ's emptying out or voiding of his divinity as he becomes incarnate in flesh. Malabou notes the way in which for Hegel kenosis is the movement by which God, on becoming incarnate, places himself outside of himself and alienates himself from himself, and thus accomplishes his divine essence in the contingent body of Christ. This movement exactly mirrors the dialectical structure of Hegelian subjectivity and its transcendental origin, or, as Malabou puts it: "There is an essential and indissoluble rapport between the kenosis of the divine and the emptiness of the transcendental. Working his way from one void to another, Hegel brings to light the kenotic essence of modern subjectivity" (1996, 155; 2005a, 111). Thus, transcendental subjectivity, in its plasticity and dialectical becoming, is from the outset characterized as vacuity, as void.

Given the Hegelian context, none of this should be in any way surprising, but it is important to trace the way in which the central place given to the void in Malabou's understanding of plasticity develops in subsequent philosophical contexts and engagements. So in *The Heidegger Change*, plasticity is recast in terms of an originary ontological mutability and transformability. Malabou demonstrates the way in which Heidegger's thinking of ontological difference and his destruction or overcoming of metaphysics testifies to the convertability of different regimes of being and to an understanding of being itself as a primordial economy of change, modification, or exchangeability. What Heidegger's thinking shows us therefore is that "being is nothing . . . but its mutability, and that ontology is therefore the name of an originary migratory and metamorphic tendency" (2004a, 344; 2011, 270). Yet if being is nothing but its mutability, nothing but originary exchange, change, and transformation, then it is, in an important sense, also and simply nothing in and of itself. It cannot be named as substance and is necessarily without essence or ground. In the place of ground or substantial foundation, there is only a void of metaphysical groundlessness. It is this groundlessness that informs Malabou's thinking of the fantastic in philosophy that gives the subtitle to *The Heidegger Change* (that is, *Du Fantastique en philosophie*). The originary ontological groundlessness or void makes it necessary that the site of being is not nameable as such by means of anything other than phantasmatic images, or as Malabou puts it: "The fantastic: the locus of originary (ex)change can only ever be invested with images. The concept falls forever short of it" (2004a, 95; 2011, 71). So this fantastical character of ontology, the nameability of being only in the plastic and mutable phantasmatic image, means that originary being only shows itself in and as a series of masks, each succeeding the other, but nowhere in this succession does anything lie behind or beneath the mask itself. Originary exchangeability, change, and transformation are as such only in and through this primordial absence or void of essence, substance, or ground.

These considerations may seem to be abstract, and one might wonder whether being in Heidegger, like the Hegelian kenotic subjectivity, thought as an emptying out or absence of substance, can ever be understood as materiality, as the very real malleability and plasticity of material forms, let alone be linked to the concerns of neuroscience. Yet Malabou takes pains to point out that the exchangeability of being that she is trying to discern as the most originary economy of Heidegger's ontology always concerns the material real—our own material real and its capacity to change, to manifest itself differently in the destruction of old, and the emergence of new, material forms. Here the originary void, the transformability of phantasmatic images of being and the plasticity of material forms, beings, and entities, all need to be thought together.

Malabou's conjugation of the biological phenomenon of neuroplasticity with the philosophical conceptions of plasticity and transformation drawn from Hegel and Heidegger achieves just such a thinking together. As was seen earlier in *What Should We Do with Our Brain?* (2004b, 2008), the dialectical, plastic subjectivity described by way of Hegel in the earlier work is recast as the plasticity of the neuronal self—the self that emerges, and that is articulated in and through, the formation of neural connections and networks within the brain. Here brain plasticity is that capacity of neural networks to receive and to give form, to be formed, deformed, and reformed throughout the life of the brain in its ongoing interactions with its surrounding environment and world. The neuronal self, it will be recalled, is mapped only in the material structure of its neural connections and networks, and according to the properties of lifelong neuroplasticity, it is structured by the dialectical play of the emergence and annihilation of form. Yet for all its biologically determined physicality, the neuronal self here is no less ungrounded and exposed to a void of substance or being than Hegel's kenotic subjectivity or Heidegger's originary economy of ontological mutability and exchange.

This is the point, as was indicated earlier, where things become particularly difficult and complicated in Malabou's thinking, for, to repeat, she needs to find some way of articulating the ontological continuity between what are, after all, two different registers: that of biological and neurobiological physicality on the one hand, and that of transcendental or phenomenological subjectivity on the other. Or, put another way, an ontological continuity needs to be articulated between one register that erases the transcendental and another that requires it. It is here, as was shown earlier, that Malabou nevertheless uses the language of discontinuity, of rupture, breach, and abyssal opening, and it is precisely here that the place of the void can be discerned once more. For if the neuronal and the mental, in their ontological continuity, nevertheless necessarily remain as two distinct registers, then the continuity at stake here is by no means a seamless continuum. Malabou, remember, insists in *What Should We Do with Our Brain?* on the existence of an "ontological explosion" in the relation between the two registers (2004b, 147; 2008, 72), the existence of "a rupture, the violence of a gap that interrupts all continuity" (2004b, 149; 2008, 73), of "a series of leaps or gaps" (2004b, 156; 2008, 75). And as was suggested earlier, this might sound rather like a straightforward reintroduction of the mind–body split, an acknowledgment that whatever neuroscience might tell us about the undeniable complicity of the cerebral and the psychical, there will always necessarily be a gaping chasm or opening between them. Yet arguably this is not so. Rather than a breach between two distinct ontological realms—the substance of physical and biological form on the one hand and the ideality or transcendence of thought on the other—what is at play here is the groundlessness, abyssal opening, or ontological and metaphysical void that subtends both instances and that makes possible the play of transition or transformation from one to the other. The ontological groundlessness of both the neuronal and the mental indicates that neither is a self-sufficient substance in itself. Both are articulated only in their relationality, and the relation of transition, transformation, or exchange from one to the other is made possible by the material spacing, differentiation, and singularization that ontological groundlessness makes possible in the first instance. This recalls the relational structure of sense elaborated in chapter 2.

The role of ontological groundlessness in thinking the mutual intrication of the neuronal and the mental is developed more explicitly as such in Malabou's work on epigenesis and rationality. In *Before Tomorrow* (2014), epigenesis, as thought from the biological perspective, and the epigenesis of reason, as thought by Kant, are brought together in much the same way as biological neuroplasticity and Hegelian plasticity are brought together in *What Should We Do with Our Brain?* (2004b, 2008). Here it is worth recalling that within biology, neuroplasticity is a specific manifestation of epigenetic development. Epigenetics in general studies the way in which the information contained in the genome is translated into the form and behavior of a biological organism, specifically taking into account the role of environmental influences and biochemical mechanisms that affect morphological and behavioral development without changing the DNA sequence itself. Neuroplasticity must thus be understood as the development of neurons, synapses, and neural interconnectivity in general, under the influence of biochemical and environmental factors that supplement and interact with the expression of genes while leaving the gene sequences themselves intact.

The most striking and original achievement of *Before Tomorrow* is to argue that the epigenetic paradigm needs to be considered as a new form of the transcendental. In a complex and wide-ranging reading of Kant's thinking about the epigenesis of reason in the first *Critique*, Malabou argues that the transcendental moment of thought should no longer be understood in terms of a strict a priori of logical invariance or predisposition toward logical or categorical structure. Rather, the a priori structure of thought must be understood as being folded into the temporal and material becoming of epigenetic development, and specifically that of neuroplasticity. But what this means, of course, is that the logical and categorical structure of thought, as folded into epigenetic becoming, formation, and articulation, has no timeless, transcendent, or fixed essence or ground. Once again, thought is shown to unfold in and as an absence of ground, or as Malabou puts it: "There is an epigenesis of reason because the a priori has no meaning. Rationality engenders itself—invents its forms—out of this necessary lack" (2014, 169; 2016, 98). The absence of a timeless or transcendent ground within thought is central to the thinking of the epigenetic transcendental in *Before Tomorrow* and is explicitly affirmed on a number of occasions. So epigenesis is, for instance, "the origin born of the lack of origin, the lack of meaning of the origin" (2014, 170; 2016, 98). Yet this is not to be understood in privative terms since "the absence of a foundation is a resource and not a lack," and indeed, "it is perhaps the resource of absence that defines the transcendental" (2014, 180–81; 2016, 105), leading finally to the conclusion that "the transcendental is not based on anything" (2014, 271; 2016, 157). Malabou's singular achievement here is to argue that an absence of foundation of ground is by no means an absence of reason, and that on the basis of an ontologically groundless biological epigenetic process, reason can find its form.

Arguably the conjugation of the neuronal and the mental that was first elaborated in *What Should We Do with Our Brain?* finds here its most philosophically developed and accomplished form. The bringing together of neuroplasticity and Hegelian subjectivity is superseded by the coarticulation of a biological and philosophical reinterpretation of epigenesis and of epigenetic temporal becoming. Once again, the ontological void finds its place, here as the most originary condition or resource of epigenetic plasticity, allowing epigenesis to be reformulated as the transcendental moment within a strictly biological account of the origin of thought and reason.

In this context, the ontological explosion between the neuronal and the mental, the series of leaps or gaps between them and across which they relate to each other are, remember, not a traditional mind–body gap but rather the void of substance that articulates relational being as such. The relation and transition of the neuronal to the mental is made possible by the fact the biological sense relations are just that—relations of sense or meaning, and not substances or instances of simple matter to which thought can be reduced. Relationality here is the material spacing, genesis, and epigenesis of biological structure (sense in Nancy's and Canguilhem's terms, sign relations in the language of biosemiotics) and occurs always and only in and as the absence of substance within structure, as the void, as the absence of origin or ground within material existence.

The epigenetic transcendental, Malabou writes, must be posed as the question of whether it is possible to speak of the transcendental in general, not as logical invariance but rather as something like a "hermeneutic latitude, the power of *sense*, opened in the heart of the biological?" (2014, 153; 2016, 89). In asking this question in this way, Malabou has laid down a major challenge to all neurobiological thought that is committed to reductivism and eliminativism and to the rejection of a naturalized transcendental moment within thought itself.[4] She has shown that it is possible, via the notion of the epigenetic transcendental, to inscribe an order of sense into both the biological ordering of organisms and in the environmental relationality and reflexive self-relationality that articulates consciousness as such. In this respect, what can now be called her epigenetic naturalism is extremely close to the speculative naturalism elaborated in relation to Nancy and Canguilhem in chapter 2. Like Nancy, Malabou has also shown that the void necessarily has its place in both the neurobiological and the philosophical perspectives as well as in the possibility of bringing the two together. Ultimately, perhaps, the ontological void has no specific place but rather is the possibility of the emergence of any material site, space, or place as such, of the relational spacing and differentiation, formation, and transformation of the real in the fundamental plasticity of its material becoming.

Organology and the Noetic Brain

These considerations can be carried over, albeit in a different manner and a different philosophical register, into Stiegler's philosophy, where his materialism and naturalism need to be thought not under the conjoined motifs of plasticity and epigenesis but rather with the context of his organological framework and under the motif of the technique of thought.[5] Since the publication of the first volume of *Technics and Time* well over twenty years ago, Stiegler has shown us, perhaps more than any other thinker discussed in this book, the extent to which the technique of thought is not just a question of the style or experimental technique of philosophical thinking. Rather, it refers also to the fundamental technicity of thought itself and the way in which technical traces condition the milieu that constitutes thought as such. In Stiegler, the technique of thought points us first and foremost to the understanding of thought itself as technical through and through. When written in French, "the technique of thought" reads "la technique de la pensée" and can refer both to the technique of thought understood as its methods, protocols, or procedures as well as to its fundamental technicity. It affirms the thoroughgoing materiality of thought, of consciousness, of reason, and of all mental activity of any kind both individual and collective.

It is in this context that the relation of thinking and embodiment in Stiegler needs to be understood, and with this the relation of his philosophy in general to the sciences and to scientific knowledge. Stiegler regularly refers to the activities of thought, consciousness, and of both subjective and intersubjective experience as noetic, a term derived from the Greek *noesis* and *nous* (via Aristotle) and referring to the mind and to the understanding or awareness that is the condition of human rationality. The question here would therefore be the relation between the noetic (mind), the material field of technicity, and the cerebral. Stiegler speaks of the noetic brain as an instance that, although biological and physiological through and through, is also the site of individual and collective phenomenal consciousness or experience. In this context, he synthesizes, in a manner similar to yet different from Malabou, postphenomenological thinking with, among others, psychosocial, anthropological, biological, and neuroscientific frameworks. In so doing, he reconfigures the relation of philosophy to science within the context of what one might call an organological requalification of knowledge in general. Indeed, in some of his most recent works, Stiegler has explicitly called for such a requalification, and in a manner that refers to both his long-standing and more recent engagements with scientific thinkers who have thought technics and life together as co-originary. For example, in the first volume of *The Automatic Society*, he speaks of "an urgent need to redefine the noetic fact in totality—that is, in every field of knowledge (of how to live, do, and conceptualize)—and to do so by integrating the perspectives of André Leroi-Gourhan and Georges Canguilhem, who were the first to posit the artificialization of life as the starting point of hominization" (2015a, 24; 2016a, 9). Stiegler also explicitly refers in this context to Gilbert Simondon, a thinker who in a certain way forms a bridge between his early interest in Leroi-Gourhan and the more recent central preoccupation with Canguilhem. Here it is Simondon's thinking of technics and individuation that will provide the impetus for a requalification of the entire field of thought and knowledge: "The Simondonian perspective on the human and the machine means that the human, as *psychic individual*, and the machine, as *technical individual*, must *constitute a new relation* where thought, art, philosophy, science, law and politics must form *a new understanding of their technical condition*" (2015a, 106; 2016a, 56). This imperative to rethink or requalify the field of knowledge according to its organological conditioning is thus entirely consistent with the Stieglerian image of philosophy that was identified in chapter 1; it does not just have an epistemological force but also has implications for the way in which philosophical thought and political action within the field of culture will relate each to the other. Again, Stiegler is explicit on this point: "*What is absolutely correct* is to posit that today the question of *organological invention* constitutes the *categorical imperative* common to all struggles—juridical, philosophical, scientific, artistic, political and economic—common to all those struggles that must be carried out against this state of fact and for a state of law, and where this is not a matter of resisting, but of inventing" (2015a, 117; 2016a, 62). In Stiegler's most recent work, then, the overcoming of the forgetting of technics that he originally identified both at the origin of Western philosophy and at the inception of the human as such becomes a twofold task: that of remapping the field of knowledge in general and of leading a struggle within wider culture for a renewed form of knowledge. This will be a struggle for a knowledge that is no longer subordinated to the order of technocratic calculability and the ends of technoscientifically dominated social, economic, and political forms.

Stiegler's inventive use of key scientific terms such as entropy, negentropy, and the Anthropocene in his most recent work needs to be understood in this wider context of the organological requalification of knowledge on the one hand and the struggle to reinvent our present and future epochs in the light of such a transformed image and practice of knowledge on the other. It is in this context that his recent discourse articulates a passage from a strictly scientific use of terms such as entropy and negentropy to a use of these terms that exceeds their scientific meaning insofar as they come to refer to questions of value within biological life, and within human life in particular. In this way, "negentropy," for instance, comes to articulate or refer to a positive valuation and affirmation of life and is incorporated into a critical-philosophical discourse on the Anthropocene that yields other invented terms such as "neganthropology" or the "Neganthropocene." These have a psychosocial or political-cultural rather than a strictly scientific sense. Similarly, it is in the light of such considerations that Stiegler's recent appeals to neuroscience and what might generally be termed his organological naturalism need to be understood.

Noetic Life in the Anthropocene: Entropy, Negentropy, and Neganthropology

Stiegler's references to entropy, to its opposite, negentropy, and to the wider scientific context of thermodynamics are rooted in the claim that, with the exception of Bergson perhaps, "the implications of the law of entropy brought to light by thermodynamics [. . .] have not been understood by so-called 'continental' philosophy, and in particular by 'French' philosophy" (Stiegler 2016b, 374). It is therefore worth noting that the overall functioning of these terms relates as much, if not more, to their philosophical interpretation than to their strictly scientific use within the context of thermodynamics, and the study of the physical systems such as they are subject to its second law in particular. As Stiegler puts it explicitly in the introduction to *The Automatic Society*: "The theory of entropy succeeds in redefining the question of *value*, if it is true that the *entropy/*

negentropy relation is the vital question par excellence" (2015a, 25–26; 2016a, 10). When he uses these terms, therefore, he is talking less about the properties of matter, of heat and energy within the systems studied by physics, and far more about a specific energetics of life and the manner in which the question of value in relation to life can be posed in this context. What is at stake here are the philosophical consequences of the assimilation of thermodynamics into the biological questioning of life. The key points of reference Stiegler draws on are Schrödinger's classic short work *What Is Life?* ([1944] 1962) and recent research inspired by Schrödinger concerning biological organization and negative entropy by Giuseppe Longo and Francis Bailly (2009).

It will be remembered from the discussion of Nick Lane's *The Vital Question* (2015) in chapter 2 that debates about the status and nature of life are still very much ongoing and open ended within contemporary biology and biological thought (Gayon 2010). Lane notes the dominant emphasis played by the DNA code and the concept of information in the biological understanding of life throughout the later twentieth century and into the twenty-first, but he argues for a greater recognition of the role played by redox chemistry, ATP synthase, and the promotion of the self-organization of matter by the flux and flows of energy across membranes or gradients (2015, 65, 73, 94). In his short discussion of Schrödinger's *What is Life?*, however, Lane does note that the great physicist emphasized both the role played by information and the genetic code, and the notion that life can be defined as that which "resists entropy, the tendency to decay" (51). It is this idea of life as a resistance to decay and to the resolution into a state of equilibrium (i.e., death) that gives rise to the notion of negative entropy. For his part, Schrödinger originally defined negative entropy in the following terms: "A living organism continually increases its entropy—or as you may say, produces positive entropy—and thus tends to approach the dangerous state of maximum entropy which is death. It can only keep aloof from it, i.e., alive, by continually drawing from its environment negative entropy—which is something very positive [. . .] what an organism feeds upon is negative entropy" (Schrödinger [1944] 1962, 25). So a living organism, as organized matter, is constantly tending toward the disorganization that results from decay (the movement toward death, apoptosis, and so on) and only fends this off by drawing energy from its environment in order to maintain its structured and complex organization for as long as the limits of the biochemical processes of life allow. In so doing, it maintains itself in a state of disequilibrium but thereby also necessarily increases the entropy in its surrounding environment.

This view, taken over by Lane and by Stiegler, is broadly in line with the widely accepted contemporary understanding of living organisms as open systems that are able to maintain themselves in far-from-equilibrium states on the basis of energy flows that allow for the building and sustaining of complex organizational structures (with the additional idea that such structures can be reproduced and can evolve over time on the basis of the transmission of genetic code).[6] For their part, Longo and Bailly note in "Biological Organization and Anti-entropy" (an early version of which is cited by Stiegler) that the concept of negative entropy understood as a principle of biological organization is not a concept that is operative within physics or the thermodynamics of nonbiological physical systems (2009, 63). Their concern is to offer a (mathematically) formalized framework that would allow them to specify the exact role played by negative entropy in living organisms, and in particular to determine the correlation of negative entropy with order, organization, and biological complexity (Longo and Montévil 2014, 215–48).

Within the context of Schrödinger's seminal short work *What Is Life?*, within Longo's (Longo and Montévil 2014) recent mathematical formalization of biological organization, and within wider contemporary and scientific acceptations of life more generally, the notion of negative entropy, or what Stiegler will call negentropy, is well established. The question here is how Stiegler comes to assimilate this established scientific use into his philosophical discourse and how it comes to be interpreted within a Stieglerian organological framework. In *The Automatic Society* (2015a, 2016a), Stiegler offers a vision of the way the advent of thermodynamic machines during the industrial revolution may have transformed our vision of the cosmos. The vision he offers is entirely consistent with his view that thought is always, in one way or another, conditioned by the technical apparatuses and systems through which we form a sense of the world by retaining a past and projecting a future. Stiegler notes that the image of the cosmos offered by Western philosophy at its beginnings was one of equilibrium, of order and harmony (echoing Nancy's characterization of the cosmos that was discussed in chapter 2). In this context, he argues, the equilibrium of an eternal cosmos is opposed to the disequilibrium of the corruptible flesh of mortals, and human technicity is marginalized or suppressed as a contingent element that has no primary status in relation to the balanced order of nature (2015a, 26; 2016a, 10). It is clear that Stiegler here is replaying his original thesis about the repression of technics at the origin of Western philosophy in terms of equilibrium/disequilibrium, order/disorder, and nature/culture (i.e., technics). With the advent of the Anthropocene, the epoch where human activity has begun to have a significant global impact on the earth's geology and ecosystems (more about this shortly), the whole notion of the cosmos as a stable, ordered equilibrium, Stiegler argues, has had to be abandoned in favor of a different vision. He links this specifically to

> the advent of the thermodynamic *machine*, which reveals the human world as being one of *fundamental disruption*, inscribes processuality, the irreversibility of becoming and the instability of equilibrium in which all this consists, at the heart of physics itself. All *principles* of thought as well as action are thereby overturned.
>
> The thermodynamic machine, which posits in *physics* the new, specific problem of the dissipation of energy, is also an industrial technical object that fundamentally disrupts *social* organizations, thereby radically altering "the understanding that being there has of its being." (2015a, 26–27; 2016a, 10–11)

Just as the science of thermodynamics determines the relations between heat and energy or work and is itself derived from the study of modern engines, so these engines themselves, Stiegler argues, have transformed the actuality of our social, economic, and political relations as well as the overall human relation to nature, and with this our image and understanding of thought and being in general. We can no longer think of an ordered and balanced eternal cosmos that sits in opposition to the disordered flux of human mortality and its contingent technical supplements. The irreversible flows of energy, dissipation, exchange, and the inexorable directionality of the movement from order to disorder now underpin the entirety of our thinking about physical and biological systems, but also about human social, cultural, economic, and political processes.

This sweeping vision in which the advent of thermodynamic machines and of the science of thermodynamics itself has overturned "all the principles of thought" may seem far too sweeping for some. But as has been suggested, it is entirely consistent with Stiegler's organological perspective and the demand that has been characterized here as an organological requalification of knowledge. Most importantly, it allows for the notions of entropy and negentropy to have different but homologous meanings as well as a heuristic value when applied to different orders or levels of knowledge and understanding: those of physics and of physical systems on one level, of biology and of biological organization on another (two strictly scientific orders of discourse), but also on further levels or layers: those of noetic, cultural, social, political, and economic life. This is not a framework that collapses the scientific into the nonscientific in an abusive or incoherent manner, but rather one that allows for the emergence of a more holistic philosophical vision in which the concept of energy "constitutes *the matrix of the thought of life as well as information, and as the play of entropy and negentropy*" (Stiegler 2015a, 28; 2016a, 11).

What is at stake here is a thinking of life in general that incorporates or brings together the physical and the biological with the noetic, cultural, social, and political. However, even if one accepts that Stiegler's use of the terms entropy and negentropy implies different orders or layers in which their strictly scientific meanings in physics and biology give way to transformed meanings that are homologously describing processes proper to the collective nonbiological organization of human life, one might still wonder how the concept of energy is coherently operating on these different nonscientific levels or layers. A skeptical reader might question whether scientific vocabulary is not being called on to think about or interpret psychosocially grounded processes that really have nothing in common with the scientific theory (in this case thermodynamics) in which such vocabulary has its properly rigorous and meaningful place. Yet as any student of mid- to late twentieth-century French philosophy knows well, the tendency to understand human signifying practices in terms of what one might call an energetics of thought is well established within this tradition. From at least the early 2000s onward, Stiegler has explicitly situated his critique of contemporary capitalism within the framework of the thinking of libidinal economy derived from the assimilation of Nietzschean thought into French philosophy by thinkers such as Gilles Deleuze, Pierre Klossowski, and Jean-François Lyotard in the 1960s and 1970s in particular (albeit in a novel form developed well beyond the scope of Lyotard's understanding of libidinal economy). This all-too-familiar context for those engaged in the study of recent and contemporary French thought is worth emphasizing because it allows the transition Stiegler makes from using the terms "entropy" and "negentropy" in the strictly scientific sense to deploying them according to a different psychosocial meaning to be more precisely specified.

In this context, to speak of thought and meaning variably in terms of energy, intensity, or relations of force implies a philosophical framework of value or evaluation that associates high energy, greater intensity, or active forces with the affirmation of singularity and therefore also with the affirmation of life and of singular living beings in general. Conversely, low energy, lesser intensity, or reactive/passive forces are associated with the affirmation of gregariousness, the evening out into sameness, and the diminution or negation of life. It is this Nietzschean-derived schema of evaluation onto which Stiegler maps his psychosocial usage of the terms "entropy" and "negentropy" as his discourse shifts from a purely biological thinking of life processes to a cultural and anthropological interpretation of our contemporary epoch and historical becoming. It is in this context that his sustained recent engagement with the notion of the Anthropocene needs also to be understood.

The Anthropocene, a term originally coined by Paul Crutzen and Eugene Stoermer in 2000, today remains subject to widespread interpretation and debate across disciplines. In its barest form, it refers to the geological epoch in which we currently find ourselves and which has only recently succeeded the Holocene. It designates the epoch in which humans have begun to have a significant and lasting impact on the earth's geological record and ecosystems. As was widely reported at the time, in 2016, members of the Working Group on the Anthropocene (WGA) declared at the Thirty-Fifth International Geological Congress in Cape Town that the term should officially be used to describe the current geological epoch and that its beginning should be dated from around 1950 with the advent of widespread nuclear testing.[7] On the basis of this declaration, members of the WGA set out to establish further scientific evidence so that the Anthropocene can become the official name for our current geological era. If it does become firmly and officially established within geological and other sciences, the Anthropocene, in its strict technical sense, will mark the era in which radioactive trace elements, anthropogenic climate change, nondegradable substances such as plastics, and many other effects of postindustrial and technoscientific human activity have all left their lasting mark on the earth's geological record.

As was indicated earlier, in *The Automatic Society* (2015a, 2016a), Stiegler links the inauguration of the Anthropocene with the advent of thermodynamic machines and their wide-ranging impact on human social, political, and economic organization as well as on our understanding of nature, being, and existence more generally. Insofar as the Anthropocene is a designation that takes geological, ecological, and anthropological realities and effects into account, it is not at all surprising that it has recently become a key point of reference in Stiegler's organological discourse. For Stiegler, the Anthropocene does not simply designate the various impacts of human activity on earth systems but also describes the technoscientific character of our postindustrial era, and more specifically the manner in which human thought, knowledge, and culture are organized by technoscience within and by that era. It describes the way in which the vast output of modern technical production has not only had an impact on the earth but has also transformed human reason insofar as scientific rationality has become progressively co-opted and subverted by technoscientific forms of understanding. As Stiegler puts it in *Dans la disruption* [In disruption]: "The Anthropocene, insofar as it requisitions science with a view to transforming it into technology [. . .] ends up structurally short-circuiting it and radically diverting it, that is to say, it comes to replace noetic activity [. . .] with an automatized understanding which functions without reference to reason" (2016b, 344). It is in this context that Stiegler comes to view the contemporary Anthropocene, and the human rationality produced by it, as dominated by an order of automatized calculability, a computational model of thought in which algorithms and other automated processes embedded in our technical society reduce or otherwise entirely negate the autonomy of human reason and decision. The Anthropocene, for Stiegler, does not just describe the embedding of lasting anthropogenic signals within the geological record but rather concerns the embedded value system of our technoscientific age which has engendered an endemic wasting of human reason, a "de-noetization by calculation" (2016b, 349).

It is on this basis that the terminology of entropy and negentropy in Stiegler's discourse converges with the concept of the Anthropocene to yield neologisms such as "neganthropology" and the "Neganthropocene." For, as the age in which human reason and collective life are subjected to a widespread and totalizing leveling out of experience insofar as it becomes organized according to an order of automated and computational calculability, the Anthropocene is understood by Stiegler to be an age of entropy. It is an age where, according to the post-Nietzschean schema of value mentioned above, the intensity and energy of living entities, their singularly structured consciousnesses or, as it were, their noetic singularity and difference, are dissipated and evened out into an undifferentiated field of sameness. This leads Stiegler to speak explicitly of "the entropic becoming of the Anthropocene" (2015a, 131; 2016a, 69). In this context, negentropy becomes a positive value that affirms the opposite of the leveling out of intensity, energy, singularity, and so on, leading Stiegler to speak of negentropic processes within the Anthropocene: "Negentropic, that is to say, singular and as singular, incalculable, intractable" (2016b, 51). In this way an entire scheme of philosophical evaluation is elaborated that allows Stiegler to identify and precisely specify both the reactive (entropic) forces that govern our postindustrial epoch and the potential for opposing active (negentropic) forces that need to be called on to inaugurate something different and new. The central arguments of *The Automatic Society* and *Dans la disruption* are set in the service of just such an inaugural gesture whereby we can begin to conceive of a "beyond" of the age of calculability and denoetization in which we currently stand. It is in the service of just such a gesture that Stiegler coins the term "neganthropology," a new body of knowledge that, he says, is "entirely to be elaborated": "We call *neganthropic* that human activity which is explicitly and imperatively governed [. . .] by negentropic criteria" (2015a, 32; 2016a, 14). Such a body of knowledge would mobilize negentropic forces in order to invent a new neganthropic order, it would be "*negentropy*, constituting a *neganthropy* and opening the age of the *Neganthropocene*" (2015a, 161; 2016a, 86). In this way, Stiegler argues, one can begin to conceive of a situation in which it is possible to turn "the entropic becoming of the Anthropocene into a negentropic becoming establishing the Neganthropocene" (2015a, 131; 2016a, 69). Stiegler's thinking is nothing if not ambitious for his "organological requalification" of human thought and knowledge has as its aim nothing short of the establishment of the futural epochal horizon of the Neganthropocene, a period in which the organization and values of our current era would be utterly transformed.

There are distinct echoes in this analysis of Heidegger's epochal critique of nihilism and his understanding of modern technology as "das Gestell," or the enframing that unconceals Being and reveals it as a calculable order of exploitable standing reserve (1993, 307–41). Yet whereas Heidegger famously asserted in his 1966 interview with *Der Spiegel* that "only a god can save us" and that such a god cannot be brought forth by thinking but can only be made ready for, Stiegler clearly does affirm that thinking, and his organological thinking in particular, can help usher in the Neganthropocene (Heidegger 1976, 1981).[8] However one might judge the effectivity of philosophical thought when it comes to such large questions of epochal transformation, what should nevertheless have become clear from this careful reconstruction of Stiegler's arguments is that they bring together scientific and philosophical registers into an overall interpretative framework while at the same time not entirely collapsing them into each other or rendering them indistinct. The by now semi-official geological designation of the Anthropocene (as the era succeeding the Holocene) is supplemented with a philosophical and organological interpretation of the term (as the era of technoscientific nihilism) just as the physical/biological concepts of entropy and negentropy are supplemented with nonscientific meanings that relate to the evaluation of life and its affirmation as singularity and difference.

Carefully separating out the scientific and nonscientific uses and meanings of Stiegler's terms allows for the relation of the physical/biological to the noetic to be precisely questioned. It also allows for a more precise and pointed interrogation of Stiegler's relation to naturalism. Even if one accepts that "energy" as conceived in the entropy and negentropy of biological organisms needs to be understood differently, albeit homologously, with the entropic or negentropic value of human life understood psychosocially, culturally, and historically, one might still wonder how and why these distinct meanings can and should be mapped so closely onto each other. More importantly still, one might question how and in exactly what way such different meanings and the distinct levels on which they operate can be said to be in a relation of ontological continuity with each other. For on the one hand biological entropy and negentropy describe or measure energy quantitatively insofar as it can be correlated with the processes of living organisms, their relation to their environment, and their organization and maintenance of internal complexity (as in the work of Longo and Bailly 2009). On the other hand, psychosocial or noetic entropy and negentropy refer to qualitative states of lived experience, to the way they are lived and experienced immanently, more or less intensively or singularly and on a subjective and intersubjective level that is not at all amenable to quantitative (and therefore scientific) determination. The division or difference that emerged clearly at the end of chapter 2 between the qualitative sense of lived existence (Nancy) and the quantitative and measurable being of existence understood as information (Ladyman and Ross) once again presents itself in stark terms, just as it did in the first half of this chapter and in the discussion of Malabou and the relation between the neuronal and mental or the (phenomeno)logical and the (neuro)biological.

For his part, Stiegler is quite clear that philosophical naturalism of the traditional kind (as described in the introduction) is not something that his organological thinking can ally itself with. This is because he places such naturalism squarely and firmly in relation to the entropic forces of the Anthropocene and the way in which such forces seek to subordinate all knowledge to a technoscientific order of calculation. For Stiegler, this is so because naturalism has allied itself with cognitivism, and cognitivism in turn has allied itself with the computational model of thought and with an algorithmic rationality and mode of governmentality: "The *computational model* of this algorithmic governmentality combines with the *naturalistic model* of current cognitivism, so that noetic life as well as biological life is reduced to a calculation" (2015a, 215–16; 2016a, 119).[9] If a distinctly Stieglerian or organological naturalism is to be elaborated, it will not only have to demonstrate some kind of relation of ontological continuity between the (quantitative) scientific and (qualitative) nonscientific registers he maps so closely each onto the other. It will also have to distinguish itself decisively from cognitivism and the manner in which it inserts itself into the wider logic of knowledge as calculability that governs the entropic becoming of the Anthropocene such as Stiegler conceives of it. Another of Stiegler's major recent works, *States of Shock*, offers a rich resource for just such an elaboration of organological naturalism.

The Technique of Thought

In *States of Shock* (2012, 2015b), Stiegler once again brings together different registers of knowledge in what might be called transformative combination of discourses.[10] Here he integrates the earlier phenomenological, postphenomenological, and bioanthropological combinations of *Technics and Time* with, among other things, a much more highly developed Simondonian framework and a now much less marginal or secondary reference to the brain, to cerebral plasticity, and to the nervous system. It is within this novel integration or synthesis of a postphenomenological thought of technical retention, a Simondonian thought of psychosocial individuation, and biological and neuroscientific frames that something like a Stieglerian or organological naturalism can be most clearly discerned.

States of Shock is articulated around a specific integration of the discourse of neurobiology and of cerebral plasticity with the postphenomenological language of Husserlian tertiary retention or of technical traces, a post-Freudian or post-Nietzschean reference to libidinal economy, and the Simondonian thinking of a preindividual, psychosocial field and of psychic and collective individuation. The language Stiegler uses to articulate the relation of the neurobiological, the postphenomenological, the libidinal, and the psychosocial is, in the first instance at least, one of a layering of these different elements, and of the conditioning, constitution, formation, or construction of the psychic apparatus on the basis of this layering. Specifically Stiegler calls upon the language of base, or *fonds*, in French, as in the following: "Technical traces [. . .] are the *milieu* of that cerebral plasticity on the basis of which the psychic apparatus is formed, or what Simondon called the psychic individual" (2012, 21; 2015b, 8); or in the following: "From out of *psychosocial* preindividual funds [*fonds*], psychic individuation *and* collective individuation are *simultaneously* arranged, according to Simondon, and where all this presupposes, as I feel compelled to add at this point, *technical* individuation" (2012, 94; 2015b, 54). The brain and nervous system are therefore a neurobiological base on which the psyche is formed in relation to the technical milieu. At the same time, the layer of the psychosocial would be situated somewhere between that neurobiological base and the psyche, and is articulated within the milieu of technical traces itself taken as the ground of individuation, forming, conditioning, or constituting it in a dynamic becoming. Elsewhere Stiegler speaks of the *Bildung*, or formation of reason, and of individual and collective memory conditioned by technical retentions, or by the plane or field of technical hypomnesic supports. Elsewhere again he speaks of mental images as being constituted on a specific biological base or vital modality that would prefigure forms of sensory and motor perception. In all these examples, the reference appears to be to a layered model articulated as a biological base, a psychosocial and technical preindividual field or plane, and the formation or constitution of individual and collective psyche, reason, and memory out of these preceding layers.

This layered model might initially suggest some kind of reductionism whereby the neurobiological base determines psychic individuation and noetic activity in general in such a way that these can always be causally reduced or related back to that base. Yet Stiegler takes great care to argue that these layers form a dynamic system of interaction. Here the plasticity of the brain, its ability to make new synaptic connections and pathways throughout life and mold or remold itself on the basis of interaction with the external world, means that the neurobiological layer is never a static or a purely causal base. Speaking in a footnote in *States of Shock* of the becoming of the technical layer of tertiary retentions, Stiegler underlines that "cerebral plasticity forms a system with this becoming—as modification of the vital layer that, trans-formed by the technical layer, becomes the psychic layer in Simondon's sense. Here, the technical plasticity of clay, for the one who shapes it and who at the same time shapes themselves by producing knowledge, is not just a metaphor for neuronal plasticity, but its condition" (2012, 265n2; 2015b, 259n47). This dynamic interaction between layers is no unidirectional causal determination of a superstructure by its base but rather an interaction where both the psyche and its neurobiological foundation are readily fluid, moldable, or transformable and where the becoming of the inorganic technical field, although itself transformable as a becoming, represents a more solid layer, a layer that is more fixed or set but that is nevertheless susceptible to being worked on. Cerebral plasticity is thus conditioned or constituted as such by the relative solidity of the technical layer or milieu. As Stiegler puts it: "This plasticity is possible only on the condition of passing through its *sterilizing* exteriorization, that is, its *solidifying and fixing* exteriorization. [. . .] This pharmacology of the fluid and the solid is also what conditions the plasticity of the noetic brain" (2012, 189n2; 2015b, 251n27). So both the neurologically plastic brain, which is the object of empirical neuroscientific study, and the noetic brain, which thinks, reasons, and is experienced only *in* and *as* our experience of thought, are plastic, moldable, or transformable on the basis of this dynamic interaction with the relative solidity of the inorganic technical layer, plane, or field. Our exposure to, apprenticeship with, and interiorization of the technical field conditions the formation and stabilization of neural networks, of synaptic connections, pathways, and systems, which in turn lays the base for the formation of psychic individuation out of that technical field. Yet this also means that the noetic activity of the brain, of the psychical apparatus and individuated psyche, can in turn take the technical field as the object of its activity and work to transform it and influence its becoming. There would thus be some level of psychic autonomy and agency opened up here whereby the fluid or plastic psyche can turn back and work transformatively on the relatively set or solid technical field. This is the essence of Stiegler's pharmacological thinking of plastic and fluid in *States of Shock* and the condition of his call to responsibility and action in relation to the contemporary milieu of industrialized technics.

There are no grounds, therefore, to accuse Stiegler of biological reductionism or technological determinism in this complex dynamic of neurobiological, technical, and psychosocial formation. However, given the separation of each of these moments into layers, distinct fields, or planes that interact with each other, there is every reason to question the exact status of each and what this might imply for the nature of their interaction, their continuity with one another, and for Stiegler's overall attempt to synthesize them within a general organology.

The neurobiological layer or base would be first and foremost empirical and the empirical object of scientific research and knowledge. Here the brain and the nervous system appear as objects of knowledge within an already constituted world to be investigated via electroencephalograms, fMRI, and other forms of neurological scanning or experimental activity. The milieu of technical traces and retentions, as well as the preindividual, psychosocial *fonds* or ground of the psyche, cannot, insofar as they precede psychic individuation as its constitutive condition, be aligned with the empirical appearance of technical objects as such, however much these objects necessarily remain the effective means of retention and of psychosocial conditioning. Rather, the plane of the trace, of retention, and of the psychosocial ground must be in a certain Derridean sense quasi-transcendental, or more precisely, as Stiegler articulates it in *States of Shock*, it must be metaempirical: "The condition of experience that is not experienced as such by experience—which constitutes the possibility of experience, but that is still not transcendental because it itself has its provenance in experience" (2012, 255n4; 2015b, 258n28). The plane of technical retentions is very much immanent to experience insofar as empirical technical objects are the bearers of such retentions but in the same way also anterior to conscious experience insofar as it conditions psychic individuation and thus makes something like thought and consciousness possible in the first instance. So there is a sense already that this plane is irreducible to the empirical field and the kind of objectivist ontology proper to neurobiological and neuroscientific knowledge and research.

Similarly, the layer of the psychically individuated self—of thought, consciousness, or noetic activity—is also nothing objectifiable or reducible to the status of an empirical object. It is therefore clear that Stiegler's layers each have a different status: empirical, metaempirical, and psychic; or noetic and phenomenal. The question that has informed this chapter throughout is clearly reposed once again here, albeit in different, distinctively Stieglerian terms. How, in their distinctness from each other, can phenomenal and intentional consciousness or psychical life and the empirically knowable brain nevertheless exist in a continuity with each other in such a way that they can interact in the complex dynamic Stiegler describes? Clearly his transformative combination of different contexts means that the references to phenomenology in Stiegler's work from *Technics and Time* (1994, 1998) onward imply a modified or radically changed postphenomenological perspective. Yet when he talks of the noetic brain on the one hand and of cerebral plasticity, synapses, and the nervous system on the other, he is nevertheless marshaling two levels that appear to be radically discontinuous rather than being in a continuity with one another. On the one hand, the noetic brain cannot be assimilated to the same status as technical objects, for such objects are both empirical things and the bearer of metaempirical retentions or traces. On the other hand, the brain is simultaneously an empirical object known to neuroscience but at the same time the ontological site from which a world is opened as such and from which consciousness and psychical or noetic life emerge. Yet the empirical and neuroscientific study of the brain as object can know nothing of its lived phenomenological experience, except perhaps by inferring correlations with reported noetic activity or experience (with all the philosophical questions this raises, as was discussed above in relation to the question of correlation-based studies within social neuroscience). Similarly intentional consciousness and psychic or noetic activity can know nothing of the immanent materiality of its own neurological or cerebral base; it can only approach brains objectively or from the outside, taking them as objects to be imaged or explored within an empirical field by technical apparatus.

Between the neurobiological, plastic brain, empirically known, and the noetic, psychically, and phenomenologically or ontologically disclosive brain, a gap or void appears to open up once more (recalling the leaps and gaps that separated the mental from the neuronal in Malabou). The two are in some kind of continuity with each other; they interact and form a system with technics, but they are also irreducible to each other.

This simultaneous continuity and discontinuity of the neurobiological and the noetic brain can best be understood in terms of an experience of, and passage to, the limits of the different perspectives and discourses that Stiegler uses to articulate the specificity of each of his distinct layers. The objectivist ontology and empirical approach of neuroscience dictate that it can only encounter the brain as a thing and observe it either directly or via the screen displays of fMRI scanners, the readings of EEGs, and so on. Neuroscience encounters a limit beyond which it cannot pass when it is a question of immanently lived consciousness and the experience of a world and of being, a dimension of which it can know nothing except by way of inferring correlations with reported noetic activities or states.

To recall once more the discussion of Malabou and fMRI studies, correlations are always necessarily supported by inferences, intuitions, and interpretative or theoretical frameworks that may be operating to different degrees of scientificity with equally variable empirical foundations. By the same token, the phenomenological or psychosocial experience of noetic activity encounters a similar limit when it seeks to think or present the lived immanent materiality of its own neurobiological base. This can be presented only as an external thing in the world accessible to the observations of neuroscience and its imaging techniques; the brain cannot experience its own materiality as such (as it might, say, feel the materiality of a hand or other exterior body part) since this is the immanent condition and cause of its possibility of experience.

At the limit of each of these perspectives—the objective-empirical and the subjective experiential or phenomenal—the gap or void that opens up is one of the impossibility of determining or making present a dimension of lived material immanence. This on the one hand would be the qualitatively lived immanence of consciousness for neuroscience, the inside of subjective experience of which neuroscience can know nothing. On the other hand, this lived immanence would be the cerebral materiality that produces the noetic activity of consciousness but that consciousness cannot spontaneously perceive or grasp as such (short of exteriorizing it into the objectified images of brain scans). In each case, and at their limits, both the objectified determinations of neuroscience and the noetic experience of consciousness encounter at their limits a lived immanence that cannot be known or determined as such. Neither neuroscience nor consciousness can present or encounter the metaempirical field of technical traces and retentions since it again is the immanent condition of both and not the simple empirical appearance of tools. These different instances, although ontologically continuous, nevertheless remain epistemologically discontinuous, irreducible to any smooth or seamless epistemic whole. This invocation of a radically unknowable lived immanence recalls Laruelle's thinking of the real, just as the invocation of limits recalls Nancy's thinking of existence as ungraspable excess. There is a sense here in which these distinct aspects of Laruellian and Nancean thought (immanence and limit-experience) combine as a means of if not resolving then at least circumscribing the mystery of the relation between the noetic brain and its physical or neurological base.

At one point in *States of Shock*, Stiegler suggests (strongly echoing Nancy) that philosophy consists perhaps always and first in thinking passages to the limit (2012, 156–57; 2015b, 93). His layering of the neurobiological, the plane of psychosocial-technical, and the psychic or noetic enacts just such a passage to the limit in respect to each. From the perspective of thought, these layers appear epistemologically discontinuous or distinct. It is only at or within that point of lived material immanence, radically inaccessible to thought and presentation as such, that the mysterious continuity of the neurobiological, the psychosocial-technical, and the noetic can be situated.

The path that this discussion has followed along Malabou's and Stiegler's different treatments of thinking bodies has led from cerebral plasticity and the creation of a novel epigenetic paradigm, through to considerations of energy and entropy in the physical, biological, and cultural spheres, and finally to a questioning of the way in which the physiological brain, technical-material traces, and noetic life are articulated in a complex relation of ontological continuity and epistemological discontinuity. In Malabou, plasticity and the paradigm of the epigenetic transcendental have as their key moment an experience of ontological void (recalling Nancy's thinking of relational sense). In Stiegler, the encounter with radical immanence at the limits of both philosophical thought and empirical knowledge offers a means of affirming ontological continuity while confirming an irreducible epistemological discontinuity. In each case, thought and matter, the qualitative and the quantitative, are once again both thought and brought together, but in a manner that defies or exceeds any logic of grounding or foundation and any horizon of unity or totality.

By way of conclusion, it might be said that what has happened in the experiments of thought that can be discerned in Malabou and Stiegler is that both philosophy and science are subjected to a specific limitation. Both philosophy and science experience their absence of ontological ground and encounter their limit (in a sense that exactly recalls the experience of limits discussed in relation to Nancy in chapter 1 and to Nancy, Smolin, and Unger in chapter 2). The experience of ontological void and of epistemological limits marks the frontier of all thought in relation to a radically unpresentable lived immanence. Philosophy as an experience of the void or passage to these limits is clearly no longer a metaphysical philosophy of foundations, grounds, and universal or created concepts. Similarly, however, science, while preserving its realism along with the specificity and validity of its methods and results, encounters the limit of its objectivist approach, its inability to determine the dimension of an immanently lived and qualitative experience that is unobjectifiable and unquantifiable as such. As was the case at the end of the discussions of both Nancy and Laruelle, this imposition of a limitation or certain modesty on science would nevertheless be a severe curtailment of the ambitions of scientism and its pretension to exhaust all knowledge of the human. The different techniques of thought developed by Malabou and Stiegler do indeed imply a naturalist continuity between the neurological and the phenomenal and among the biological, the technical-material, and the noetic. Both take thinking beyond the impasses of reductivism and eliminativism on the one hand and of dualism on the other. Yet as the experience of the void or of limits, and therefore of a discontinuity of knowledge within which that continuity must be located, this naturalism radically displaces the privilege of an exclusively scientific epistemology in its power to exhaustively determine the immanent, lived real of experience. In this way, the plasticity or technicity of thought once again articulates a decisive shift and renewal of the relation of philosophy to science.

CONCLUSION

The Eclipse of Totality

ONE OF THE KEY QUESTIONS posed by the encounters with philosophy, nonphilosophy, and science that have been staged in this book is the question of the irreducible plurality of the real. Can we admit something like a plural real while maintaining an uncompromising realism and avoiding a lapse into an arbitrary or unbridled relativism? If the real is irreducibly plural, this is not just because of its internal multiplicity and complexity but also because of the absence of any overarching metaphysical or philosophical principle that would subsume that multiplicity into a unitary foundation or ground. Or put in less Nancean and more Laruellian terms, it is because the multiplicity of the real (as One) would be indivisible and therefore always undivided by the operations of transcendence—of division, splitting, and synthesis—that would subsume it into the horizon of philosophical representation and Totality. Or the real is plural because of the infinitely groundless plasticity and mutability of its forms (Malabou), or again because its multiple forms are articulated in an originary technicity that supplements an irreducible default of the origin (Stiegler).

It was the physicist Aurélien Barrau who noted in his collaboration with Nancy that "beyond experimental results, phenomena, objects, and events, laws themselves seem to be inscribed into a plural real. Their diversity is, once again, evidently irreducible" (Nancy and Barrau 2011, 112; 2014, 63). Similarly, d'Espagnat's open realism, Dupré's promiscuous realism, and Cartwright's dappled universe or metaphysical nomological pluralism all admit a plural real while staying firmly anchored within scientific-realist perspectives. By the same token, Unger and Smolin's temporal naturalism, in its critique of mathematism and Pythagoreanism within scientific theory, and in its attempt to undo the evisceration of time from our knowledge of the universe, also affirms the multiplicity and plurality of the real. The path toward this notion of a plural real has been trodden by way of encounters with specific experiences of thought in chapter 1 and then by way of further exchanges with scientific thinking in subsequent chapters. At each stage of the discussion, the philosophers discussed have been allied with this diverse range of perspectives within contemporary science and philosophy of science debate. From the different readings given here, it has at each point been argued that the notion of a plural real can be affirmed and maintained alongside both a realism of thought and a realism of scientific knowledge.

It follows that the real, as plural, can admit a plurality of techniques by way of which it can be approached, accessed, or encountered. Again, it can admit this plurality of technique without allowing or permitting just any technique, just any mode of approach, access, or encounter. There is nothing to say that different techniques cannot be judged or evaluated for the richness and fullness of their access to or encounter with the real. It might be judged that the greater the richness and the greater the plurality, multiplicity, or complexity that can be circumscribed or presented by any one technique, the more effective its mode of approach, access, or encounter. For it is surely not unreasonable to suppose that plurality, as irreducible, can, and indeed must, be approached in multiple ways and from different perspectives, and that this will in turn yield diverse results or modes of knowledge. This diversity of modes will be no less real provided that they are, each in one way or another, determined by the real as their cause. As was argued in relation to the biological knowledge of the world discussed in chapter 2 and the realism-of-the-last-instance discussed in chapter 3, the real, whether we like it or not, and whatever our epistemic dispositions, will come to us one way or another.

It is because of this plurality of the real that the post-Continental naturalism elaborated here has emerged, not as a unitary philosophy (or nonphilosophy) but in and as a series of techniques of thought. These techniques may have similarities and countenance different degrees of consilience with each other (for example, Nancy, Malabou, and Stiegler), or they may have similarities and be irreconcilably opposed to each other (for example, Nancy and Laruelle). Yet as experiences of thought, each articulated according to their different techniques, they are all, in one way or another, responses to the specificity of certain encounters with the real. The real is encountered here in an experience of the limits of thought and of the loss of origin or ontological ground within thought. Or yet again it is experienced in a placing of thought back into a relation with radical immanence. It is, as it were, an inevitable feature of the plural real to be only really known in a necessarily nonunifiable plurality of technique.

The plurality of philosophical or nonphilosophical technique interrogated across the preceding chapters can perhaps be compared by way of analogy with the divergent approaches to the problem of quantum gravity evoked by Barrau in *What's These Worlds Coming To?* (as discussed in chapter 2; Nancy and Barrau 2011, 67; 2014, 36). In each case there is a multiplicity of theories (strings, loops, twistors) or of (non)philosophical techniques, and in each case theory or technique is a certain way of encountering, accessing, or being determined by the real as it comes to us *as* real and *from* the real, either in science (by way of experimental results) or thought (as a specific rigor and experience of thinking). Indeed, it may be the case that we need to understand the multiple emergence, development, and eventual passing into history of both scientific theories and philosophical techniques as functions of a plural real that cannot be encompassed by some definitive theory of everything or inscribed within a horizon of total knowledge or understanding.

It is in this context that the post-Continental naturalism elaborated here differs so decisively from the philosophical naturalisms of the American tradition and from more recent and contemporary scientistic and naturalist positions. The synoptic vision, total science, and total theory of Sellars, Quine, and Lewis, respectively, indicate that the American naturalist tradition was born and sustained within an image of philosophy oriented toward a horizon of totality and unity. Subsequently, the unity of science posited by Ladyman and Ross, Wallace's multiverse, Hawking and Mlodinow's ultimate theory of everything, Papineau's complete physics, and Patricia Churchland's unified grand theory of the mind–brain all maintained in one way or another this fundamental orientation toward a total vision of a scientifically knowable and exhaustible reality. The continuity between philosophy and science that forms, as it were, the backbone of the naturalist tradition serves ultimately not to subordinate philosophy to science but rather to place both equally in the service of this unifying and totalizing vision of the real and of knowledge (of the real).

The thinking elaborated here unequivocally emerges and sustains itself within an image of philosophy oriented toward an eclipse of totality and all horizons of unity and completeness. It has been argued, however, that it remains resolutely naturalist as well as realist, but only at the price of eschewing or bypassing metaphysical conceptuality or foundationalism as such. In their different, ways all the thinkers discussed roundly reject any ontological dualism of mind and body or of thought and matter. Although since the real according to which they think is irreducibly plural, discontinuities of perspective, technique, or approach remain as a necessary response to the demand made by the plural real. Where traditional naturalism may demand "an externalist approach to epistemology" (Papineau 1993, 1), these thinkers understand that, whether it is a question of internal subjective consciousness and lived immanence or of external knowledge of an objective world, our knowledge of the real, such as it is, comes to us from the real by way of real relations with the real. The opposition between internal and external or subject-dependent and subject-independent experience loses its epistemological and ontological value in this context. As regards the continuity between philosophy and science that forms the backbone of the naturalist tradition to date, it has become eminently clear that post-Continental naturalism displaces this continuity into a discontinuity between the two. Yet within this discontinuity there is a necessary and proximal entanglement of philosophy and science or a close relation of experimental and speculative exchange between the two. This entanglement and exchange are necessary because philosophy without science would sever itself from one of the most important means of accessing the real, and science without speculative, philosophical, and/or experimental thinking would lack capacity to imagine and think the real when it comes up against the limits of what it knows. Across these entanglements and exchanges, a plural series of transformations emerges; philosophy becomes nonphilosophy, or an experience of philosophy's limits, of its mutable forms, or again of the void, absence, or default of philosophy's origin.

As has also been argued at various stages throughout these discussions, this reorientation of the relation of philosophy to science has implications for the way in which we may come to understand the wider field of knowledge in general and the relation of nonscientific bodies of knowledge to the sciences in particular. Once the figure of totality has been eclipsed, the ambitions that have so often resulted from the naturalistic desire for total science or total theory (those of reductivism, eliminativism, and of scientism more generally) cease to have a heuristic or philosophical value. For instance, the discourses of the arts and humanities, insofar as they may seek to determine, in Nancy's words, "the sense of the world," can be understood as one specific set of techniques of thought according to which the plural real can be accessed or encountered (in this case in its irreducibly lived, qualitative, and symbolic dimensions). When such discourses do borrow from or interact with the sciences, they will do so nonreductively and without the prejudice that places scientific discourse in a preeminent position with regard to explanations of the human and of human experience. Once objective, quantitative, and informational knowledge is placed in an ontological continuity but an epistemological discontinuity, with (inter)subjective, qualitative, and symbolic knowledge, only then can the field of knowledge in general be construed in naturalistic terms without at the same time lapsing into a totalizing and reductive scientism.

That the naturalism elaborated here can be understood as post-Continental results not just from that fact that it brings together post-phenomenological philosophical discourse with scientific and naturalistic thinking. The term "post-Continental" also designates, or aspires toward, the (already ongoing) emergence of a field of philosophical knowledge and techniques that is pluralistic enough in its engagements to transcend the Continental–analytic divide such as it may have been. If one can imagine a pluralization of philosophical technique that nevertheless remains rigorously tied to a thinking by way of and according to the real, then such an opposition will continue to diminish in its explanatory force. By the same token, the term "post-Continental," in articulating a pluralization of philosophical technique, allows for a pluralized understanding of the sciences (according to the model of temporal and speculative naturalism described in chapter 2 or of the open sciences as described in chapter 3). This opens the way for a pluralization of techniques within the diverse disciplines of knowledge in general without sanctioning relativism or the notion that anything goes.

One might question whether it is useful or necessary to use the term "naturalism" at all in relation to the thinking of Nancy, Laruelle, Malabou, and Stiegler such as it has been negotiated, interpreted, and developed here. Perhaps what has emerged is so different from the tradition of American naturalism and its contemporary legacy that the term does not or should not apply. Yet when applied with rigor, names or labels obviously do assist emerging bodies of thought to find their place within contemporary philosophical debate. It is in the name of opening further debate around this body of thinking that the term "post-Continental naturalism" has been adopted. It is also in the name of possible further development and research that the different techniques of thought interrogated here may also be given names: speculative naturalism (Nancy), epigenetic naturalism (Malabou), and organological naturalism (Stiegler). Although using the term "naturalism" at all in relation to Laruelle might imply an all too philosophical designation, one could tentatively and provisionally refer to "axiomatic naturalism" to describe the generic science that emerges from the postulates of nonphilosophy as explored here. In using such names, paths for further research programs can be staked and possible further avenues explored.

Within the eclipse of totality that post-Continental naturalism imposes, the field of exploration across diverse areas of knowledge is therefore opened in a specific and radical manner. Although for some of the techniques of thought elaborated here an encounter with limits is decisive (e.g., Nancy, Stiegler, Malabou), what is also clear is that in the absence of any guiding figure of totality, the specific determinable limits of research and exploration cannot be prescribed or circumscribed in advance. From this perspective, science must continue as it has always done, without presupposing what its immediate limits may or may not be just as it is nevertheless obliged to proceed with greater modesty with regard to the prospective metaphysical horizons of totality, completion, or the theory of everything. In the same way, philosophical and disciplinary techniques in diverse areas may wish to demarcate their internal borders and relations to other forms of knowledge, but they will do so within the opened field of knowledge that the eclipse of totality itself opens.

The results of this ambition for an opening of thought onto a pluralization of technique also cannot be prescribed or verified in advance. Yet today the challenges of thought are such that there is an imperative to open ourselves to demands of the real that cannot be ignored. At the same time, there is no reason to assume that our existing categories and approaches are, or will remain, fit for purpose. The gulf between the scientific picture of the world and the myriad of manifest pictures and worldviews that modern technological culture supports needs to be addressed in ways that allow renewed possibilities of communication and exchange between the two to take place. The specificities of manifest pictures and of qualitatively lived experience, both individual and collective, need to be reconnected with their embeddedness in a real that makes demands. To ignore the demands of the real is to risk destruction and annihilation. Life is nothing if it is not that which resists its own destruction. If post-Continental naturalism deserves a name and a place within philosophy, science, and the field of knowledge in general, it is because its diverse techniques respond to the multiple demands made by a plural real.

NOTES

Introduction

1. For a critique of the reliance of naturalism on dualist oppositions, see Olafson (2001, 10, 16).
2. Analytic truths would be those "grounded in meanings independent of matters of fact," synthetic truths would be grounded in facts, and reductionism is the belief that meaningful statements are "equivalent to some logical construct upon terms which refer to immediate experience" (Quine 1953, 20).
3. Even the later thought of Merleau-Ponty has also often been described as naturalist and has recently been aligned with the naturalism of C. S. Peirce. See, e.g., Harney (2015).
4. Other works such as Frederick A. Olafson's *Naturalism and the Human Condition: Against Scientism* argue that the fundamental tenets of philosophical naturalism are an error and need to be corrected with reference to phenomenological and existential concepts of transcendence and being-in-the-world (Olafson 2001, 96, 105, 108).
5. For a helpful overview of this emergence, see Dolphijn and van der Tuin (2012).
6. An exception here would be the exceptionally original work of Karan Barad (2007, 436n80), which takes its distance from Deleuze and Deleuzianism.

1. The Image of Philosophy

1. For a detailed reading of Nancy's interpretation of Kant in his 1970s texts, see James (2006, 26–48).
2. For a discussion of this earlier work see James (2006, 49–63). Nancy, of course, is well aware that Kant's transcendental unity of apperception is not subjectivity as such but argues that it is implicated in the logic of foundation proper to the philosophy of the subject and philosophical modernity in general.
3. The principal recorded exchange between the two is an interview included at the end of Laruelle's 1977 work *Le Déclin de l'écriture* [The decline of writing]. In the interview, Nancy, Derrida, Sarah Kofman, and Philippe Lacoue-Labarthe question Laruelle in somewhat skeptical terms about his work.

4. Although there is not the space to make the argument here it is also arguably the case that a similar relation of philosophy to science pertains in other thinkers related to this context, e.g., Bruno Latour, and to the various bodies of thought that have emerged under the name of new materialism in the work of figures such as Jane Bennett (2009) and Rosie Braidotti (2013).
5. Laruelle offers a slight deviation from these formulations, as will no doubt already be clear.

2. The Relational Universe

1. For an account of contemporary naturalism's "disillusioned" take on reality, see Rosenberg (2014).
2. For an outline of Harman's history of speculative realism, see Harman (2011); for his most recent account of the current state of speculative realism, see Harman (2013). For other recent surveys of speculative realism, see Shaviro (2014), Gratton (2014), and Sparrow (2014).
3. Brassier's privileging of the "scientific image" of the world has more recently been nuanced (Brassier 2014, 101–14).
4. The influential work of Carlo Rovelli, one of Smolin's collaborators around theories of quantum loop gravity, offers an excellent example of this relational understanding within scientific theory; see Rovelli (1996).
5. In my first account of Nancy's thought, both its nonphenomenological character and its realism were emphasized but perhaps underplayed to the extent that both Husserlian and Heideggerian accounts of appearance were invoked as a means of presenting his thinking of the sense of the world in an introductory manner. See James (2006, 65–113).
6. Wheeler has articulated this thesis in terms of the "it from bit theory": "Every it—every particle, every field of force, even the spacetime continuum itself—derives its function, its meaning, its very existence entirely—even if in some contexts indirectly—from the apparatus-elicited answers to yes or no, questions, binary choices, *bits*" (1999, 310). However, in Wheeler's formulation, this informational existence necessarily relies on a participation of the observer with the observed phenomenon: "All things physical are information-theoretic in origin and this is a participatory universe" (311). Ladyman and Ross (2007) do not appear to commit to this strongly correlationist view of information.
7. Timpson (2013) provides a comprehensive overview of the current state of quantum information theory.
8. Wheeler's theory of existence as information-theoretic also explicitly affirms an absence of foundation insofar as he questions whether observer-participancy may be "built on 'insubstantial nothingness'?" (1999, 314).
9. Nancy's "The Heart of Things" appears in the original French as "Le cœur des choses" in *Un Pensée finie* (1990, 197–223). *Corpus* was originally written as a conference paper and appears in French in a much extended book-length version, *Corpus* (1992).
10. The original French reads: "Il n'y a donc pas un « fond » du « quelconque » le *quel-*

conque est la différence." This phrase has inexplicably been omitted from the English translation in *The Birth to Presence* (Nancy 1993a, 174).

11. On Poincaré's structural realism, see d'Espagnat (2002, 424–28).
12. In the same volume, Alain Prochiantz also criticizes the use of an information theoretic understanding in relation to the genetic code of DNA (Mathiot 1993, 271–78).
13. For a useful summary overview of this development up to 2008, see Mazzocchi (2008).
14. Understanding sense in these terms articulates a position which rejects the "bifurcation of nature" as articulated by Whitehead in *The Concept of Nature* ([1920] 2015, 21–22).
15. This account of knowledge elaborated on the basis of Nancean sense has strong echoes with Quine's grounding of epistemology in the stimulation of the sensory surfaces of biological organisms and their neural intake (1990b, 36).
16. The sense of "open" or "opening" here needs, however, to be distinguished from its sense within physics where speaking of the "open" with regard to the universe refers to geometries of negative curvature.
17. This is clearly different from Laruelle's understanding of the One, as discussed in the previous chapter.
18. Counter to this, it might be noted that the ontic structural realism defended by Ladyman and Ross (2007) is conceived specifically to respond to the problem of theory change insofar as it articulates the mathematical structure of a reality that is preserved across theory change; see Worral (1989).
19. Barrau works at the CNRS Laboratory for Subatomic Physics and Cosmology at Joseph Fourier University (Grenoble) and has been awarded a number of prestigious prizes in recognition of his research. He has published stand-alone books (2010, 2011, 2016, 2017) as well as volumes coauthored with Nancy.
20. The Stanford School was most notably associated with names such as John Dupré, Ian Hacking, and Nancy Cartwright and will be discussed at greater length in chapter 3.
21. Relational properties, Smolin argues, include "causal relations and spacetime intervals which are derivative from them" and intrinsic properties include "the dynamical quantities: energy and momenta, together with qualia" (Smolin and Unger 2015, 483). From the phenomenological perspective, understanding qualia as nonrelational is highly problematic.
22. Smolin's position on the question of the relation of life to physical processes is in fact identical to Lane's: "Living things are a particular type of process which has emerged on top of the flows of energy and cycles of materials that characterize [. . .] open systems" (1997, 154).
23. More precisely, this view can be called Pythagoreanism. Ian Hacking has noted "that the very idea of mathematical physics, that mathematics can be used to unlock the deepest secrets of Nature, owes a great deal to a Pythagorean tradition" (2014, 13).
24. Smolin's position here comes close to Nancy Cartwright's in *The Dappled World*

(1999, 2–3) when she argues that physics works by determining its objects of knowledge and the lawlike regularities that govern them within specific bounded spaces, or what she calls nomological machines.

25. A similar view has been defended by Barrau (2016).
26. It is interesting to note in relation to this point that the theories favored by Smolin (quantum loop gravity, for instance) tend to be background-independent theories—that is to say, they do not presuppose a backdrop of static parameters in the way that the Newtonian paradigm does.
27. Carlo Rovelli, who shares Smolin's relational understanding of the universe, appears to frame the question of time in a manner which is opposed to Smolin. In *Reality Is Not What It Seems* (2014), Rovelli argues that time does not exist as some kind of uniform "great cosmic clock," and that ultimately it does not exist at all (153–54). This would seem to run against Smolin's understanding of the universe as fundamentally temporal in character. However, Rovelli in fact argues that the "passing of time is intrinsic to the world, it is born of the world itself, out of the relations between quantum events which are the world and which themselves generate their own time" (154). Despite a seemingly different understanding of temporality and determinism in relation to QM, this is ultimately not too dissimilar from Smolin's view as it is explored in this chapter.
28. This broad definition of Platonism in mathematics needs to be distinguished from its more particular usage in modern mathematics, where it is largely concerned with semantic approaches.
29. For a fuller analysis of the relation between temporalization and spatialization in Nancy (and in relation to Derridean *différance*), see James (2014, 110–26).
30. The notion of speculative necessity resonates with Roger Penrose's (2016, 216–17, 323–24) arguments relating to the role of fantasy in physical theories.

3. Generic Science

1. See Laruelle (2000, 2015).
2. Poincaré (2016) limits the scope of conventions by emphasizing the degree to which truth is also grounded in empirical experience.
3. Thomas Reydon notes, "Apart from being rooted in a venerable, long-standing philosophical tradition, the idea that the furniture of the world comes in natural kinds, each of which is individuated by its own essence [. . .], does not seem to have much support for it" (2010, 177–78).
4. Howard Sankey describes scientific realism in these terms: "The world investigated by science is an objective reality that exists independently of human thought" (2004, 57). Later he adds: "The world that we inhabit and which science investigates is not an amorphous world. It is a structured world of entities, properties and relations, which fall into naturally occurring categories" (61–62). This describes what is here understood as "strong" scientific realism.
5. Dupré's ideas have been criticized elsewhere also; see Mitchell (2003) and Sklar (2003).

6. On the equality of all thoughts in Laruelle's nonphilosophy, see Ó Maoilearca (2015).
7. Wallace describes decoherence as "the process by which the environment of a system continually interacts with, and becomes entangled with, that system. Its best-known property is the suppression of coherence in coherent superpositions of states in a particular basis picked out by the system—environment interaction—hence the name" (2012, 77).
8. Wallace (2012) uses the term "branching" rather than "splitting," which is consistent, perhaps, with the notion that the multiple universes in question remain in superposition as they were before with each other rather than one universe splitting into a series of other universes.
9. On the multiplicity of Laruelle's One or immanent real, see *Le Principe de minorité* (1981, 5–6, 22–23, 111–12) and *Theory of Identities* (1992, 71; 2016, 48). That the One for Laruelle is inhabited by multiplicity may seem strange or paradoxical, but only if we fail to understand that the One of the real is One because it is undivided by the operations of transcendence. This point will be touched on again briefly in the conclusion.

4. Thinking Bodies

1. Arguably Papineau's reliance on the notion of the completeness of physics is questionable, given the debates around the unity of science that have been engaged by philosophers situated on both sides of the question, and by Ladyman and Ross (2007), d'Espagnat (2002), Dupré (1993), and Cartwright (1999). Within the context of his own discussion, Papineau affirms this by a speculative philosophical move that recalls the one identified in the introduction in relation to David Lewis's taking the many worlds hypothesis to be true. Papineau notes, "Suppose we *define* 'physics' as the science of whatever categories are needed to give full explanations for all physical effects. I accept [. . .] that this science will be different from current physical theory, and thus that we don't yet know what it is. But, even so, there is no difficulty about how we know that it is complete, for we have simply defined it so as to be complete" (1993, 29–30).
2. For an account of Malabou's thinking of freedom in this context and its political dimension, see James (2012, 95–102).
3. The paper was originally given the title "Voodoo Correlations in Social Neuroscience," but this was changed as a result of the ensuing controversy relating to its key arguments.
4. Malabou has continued with this challenge by further exploring what she calls the "milieu between biological life and symbolic life" and the extent to which "the symbolic and the biological are originarily and intimately mixed" (2017, 9, 10).
5. "Naturalism" is not a term that Stiegler would be likely to accept as a characterization of his thinking. He associates traditional naturalism, cognitivism, and so on with the logic of capitalism in the Anthropocene that he is aiming to challenge or move beyond. It is also clear that the Stieglerian thinking of technics challenges

the very notion of the "natural" and any possibility of opposing nature to culture. To this extent, in orthodox Stieglerian terms, the "organological naturalism" proposed here would be a contradiction in terms. What follows thus argues for a distinct Stieglerian naturalism and deliberately runs against the grain of what Stiegler himself might allow.

6. See, for example, Jean Gayon's synthetic overview at the end of the special journal issue on "Defining Life," where he succinctly notes the definition of life as the "individual self-maintenance and the open-ended evolution of a collection of similar entities" (2010, 242).
7. Damian Carrington, "The Anthropocene Epoch: Scientists Declare Dawn of Human-Influenced Age," *Guardian*, August 29, 2016, https://www.theguardian.com/. There is also an established scientific journal entitled *Anthropocene*, which describes itself as "an interdisciplinary journal that publishes peer-reviewed works addressing the nature, scale, and extent of interactions that people have with Earth processes and systems" (https://www.elsevier.com/journals/anthropocene/2213-3054) and which explores also how anthropogenic alterations of earth's ecosystems affect society. For a particularly useful intervention on this latter point, see Bostic and Howey (2017). Stiegler's philosophical interpretation of the Anthropocene is arguably in line with Bostic and Howey's call for a full interdisciplinary response to this new geological category and reality, one that would include the liberal arts.
8. Stiegler, however, is careful to qualify the concept of the Neganthropocene as a guiding value and ideal, or, as he puts it, as a dream: "The Neganthropocene is a noetic dream which 'in all probability' *has no chance of being realised.* It is *for that reason* that it is a dream" (2016a, 427).
9. A key point of reference here is to the work of Antoinette Rouvroy and Thomas Berns (2013). Their thesis relating to algorithmic governmentality is central to arguments of Stiegler's *The Automatic Society* (2015a, 2016a) and *Dans la Disruption* (2016b) but is beyond the scope of this discussion.
10. For a lengthy critical discussion of Stiegler's original "transformative combination" of philosophical and bioanthropological discourses see *Technics and Time 1: The Fault of Epimetheus* (1994, 1998), see James (2010).

BIBLIOGRAPHY

Ansell-Pearson, Keith. 1997. *Viroid Life: Perspectives on Nietzsche and the Transhuman Condition*. London: Routledge.

Ansell-Pearson, Keith. 1999. *Germinal Life: The Difference and Repetition of Deleuze*. London: Routledge.

Atlan, Henri. 1999. *Étincelles de hasard*. Paris: Seuil.

Bachelard, Gaston. 1934. *Le Nouvel esprit scientifique*. Paris: Presses Universitaires de France.

Barad, Karen. 2007. *Meeting the Universe Halfway: Quantum Physics and the Entanglement of Matter and Meaning*. Durham, N.C.: Duke University Press.

Barrau, Aurélien. 2010. *Multivers*. Paris: La Ville Brûle.

Barrau, Aurélien. 2011. *Forme et origine de l'univers*. Paris: Dunod.

Barrau, Aurélien. 2016. *De la vérité dans les sciences*. Paris: Dunod.

Barrau, Aurélien. 2017. *Des Univers multiples*. Paris: Dunod.

Bashour, Bana, and Hans D. Muller, eds. 2014. *Contemporary Philosophical Naturalism and Its Implications*. New York: Routledge.

Bennett, Jane. 2009. *Vibrant Matter: A Political Ecology of Things*. Durham, N.C.: Duke University Press.

Blanchot, Maurice. 1969. *L'Entretien infini*. Paris: Gallimard.

Blanchot, Maurice. 1993. *The Infinite Conversation*. Translated by Susan Hanson. Minneapolis: University of Minnesota Press.

Bonta, Mark, and John Protevi. 2004. *Deleuze and Geophilosophy: A Guide and Glossary*. Edinburgh: Edinburgh University Press.

Bostic, Heidi, and Meghan Howey. 2017. "To Address the Anthropocene, Engage the Liberal Arts." *Anthropocene* 18:105–10.

Braidotti, Rosie. 2013. *The Posthuman*. Cambridge: Polity.

Brassier, Ray. 2007. *Nihil Unbound*. Basingstoke: Palgrave Macmillan.

Brassier, Ray. 2014. "Nominalism, Naturalism, and Materialism: Sellars's Critical Ontology." In Bashour and Muller 2014, 101–13.

Butterfield, Jeremy, ed. 1999. *The Arguments of Time*. Oxford: Oxford University Press.

Butterfield, Jeremy, and Constantine Pagonis, eds. 1999. *From Physics to Philosophy*. Cambridge: Cambridge University Press.

Canguilhem, Georges. 1965. *La Connaissance de la vie*. Paris: Vrin.

Canguilhem, Georges. 1994a. *Études d'histoire et de philosophie des sciences*. Paris: Vrin.

Canguilhem, Georges. 1994b. "La Nouvelle connaissance de la vie." In Canguilhem 1994a, *335–64*.

Canguilhem, Georges. 1994c. *A Vital Rationalist: Selected Writings from Georges Canguilhem*. Edited by François Delaporte. New York: Zone.

Canguilhem, Georges. 1994d. "New Knowledge of Life." In Canguilhem 1994c, 303–21.

Canguilhem, Georges. 2008. *Knowledge of Life*. Translated by Stefanos Geroulanos and Daniela Ginsburg. New York: Fordham University Press.

Carrier, M., J. Roggenhofer, G. Küppers, and P. Blanchard, eds. 2004. *Knowledge and the World: Challenges beyond the Science Wars*. Berlin: Springer.

Carroll, Sean. 2006. "The Trouble with Physics" (review). Cosmic Variance (*Discover* blog), October 3, 2006. http://blogs.discovermagazine.com/.

Cartwright, Nancy. 1999. *The Dappled World: A Study of the Boundaries of Science*. Cambridge: Cambridge University Press.

Chalmers, David. 1996. *The Conscious Mind: In Search of a Fundamental Theory*. Oxford: Oxford University Press.

Churchland, Patricia. 1986. *Neurophilosophy: Towards a Unified Science of the Mind*. Cambridge, Mass.: MIT Press.

Churchland, Paul M. 1981. "Eliminative Materialism and the Propositional Attitudes." *Journal of Philosophy* 78 (8): 67–90.

Churchland, Paul M. 1996. "The Rediscovery of Light." *Journal of Philosphy* 93 (5): 211–28.

d'Espagnat, Bernard. 2002. *Traité de physique et de philosophie*. Paris: Fayard.

d'Espagnat, Bernard. 2012. "Physique quantique et réalité la réalité c'est quoi?" http://www.asmp.fr/fiches_academiciens/textacad/espagnat/12-05-22_Physique Quantique&Realite.pdf.

Damasio, Antonio. 1996. *Descartes' Error: Emotion, Reason and the Human Brain*. London: Papermac.

Damasio, Antonio. 2000. *The Feeling of What Happens: Body and Emotion in the Making of Consciousness*. London: Heineman.

Damasio, Antonio. 2010. *Self Comes to Mind: Constructing the Conscious Brain*. London: Heineman.

de Bestegui, Miguel. 2004. *Truth and Genesis: Philosophy as Differential Ontology*. Bloomington: Indiana University Press.

de Martino, Benedetto, John P. O'Doherty, Debajyoti Ray, Peter Bossaerts, and Colin Camerer. 2013. "In the Mind of the Market: Theory of Mind Biases Value Computation during Financial Bubbles." *Neuron* 79:1223–31.

DeLanda, Manuel. 2002. *Intensive Science and Virtual Philosophy*. London: Bloomsbury.

Delaney, C. F., Michael J. Loux, Gary Gutting, and W. David Solomon, eds. 1977. *The Synoptic Vision: Essays on the Philosophy of Wilfrid Sellars*. Notre Dame, Ind.: Notre Dame University Press.

Deleuze, Gilles, and Félix Guattari. 1991. *Qu'est-ce que la philosophie?* Paris: Minuit.

Deleuze, Gilles, and Félix Guattari. 1994. *What Is Philosophy?* Translated by Hugh Tomlinson and Graham Burchell. New York: Columbia University Press.

Dewey, John. 1951. *The Influence of Darwin on Philosophy and Other Essays*. New York: Peter Smith.

Dolphijn, Rick, and Iris van der Tuin. 2012. *New Materialisms: Interviews and Cartographies*. Ann Arbor: University of Michigan Press.

Dreyfus, Hubert L. 1982. *Husserl, Intentionality, and Cognitive Science*. Cambridge, Mass.: MIT Press.

Dupré, John. 1993. *The Disorder of Things: The Metaphysical Foundations of the Disunity of Science*. Cambridge, Mass.: Harvard University Press.

Favereau, Donald. 2008. "The Biosemiotic Turn." *Biosemiotics* 1 (5): 5–23.

Favereau, Donald. 2010. *Essential Readings in Biosemiotics: Anthology and Commentary*. Dordrecht: Springer.

Gallagher, Shaun, and Dan Zehavi. 2012. *The Phenomenological Mind*. London: Routledge.

Galloway, Alexander. 2014. *Laruelle: Against the Digital*. Minneapolis: University of Minnesota Press.

Gangle, Rocco. 2013. *François Laruelle's "Philosophies of Difference": An Introduction and Guide*. Edinburgh: Edinburgh University Press.

Gangle, Rocco, and Julius Greve. 2017. *Superpositions: Laruelle and the Humanities*. London: Rowman & Littlefield.

Gayon, Jean. 1998. "The Concept of Individuality in Canguilhem's Philosophy of Biology." *Journal of Philosophy of Biology* 31 (3): 305–25.

Gayon, Jean. 2004. "Realism and Biological Knowledge." In Carrier, Roggenhofer, Küppers, and Blanchard 2004, 171–90.

Gayon, Jean. 2010. "Defining Life: Synthesis and Conclusions." In "Defining Life," special issue of *Origins of Life and Evolution of Biospheres* 40 (2): 231–44.

Gayon, Jean, Christophe Malaterre, Michel Morange, Florence Raulin-Cerceau, and Stéphane Tirard, eds. 2010. "Defining Life," special issue of *Origins of Life and Evolution of Biospheres* 40 (2): 119–244.

Gratton, Peter. 2014. *Speculative Realism: Problems and Prospects*. London: Continuum.

Hacking, Ian. 2014. *Why Is There a Philosophy of Mathematics at All?* Cambridge: Cambridge University Press.

Harman, Graham. 2005. *Guerrilla Metaphysics*. Chicago: Open Court.

Harman, Graham. 2007. "On Vicarious Causation." In *Collapse 2: Speculative Realism*, edited by R. Mackay, 187–221. Falmouth: U.K.: Urbanomic.

Harman, Graham. 2010. "I Am Also of the Opinion that Materialism Must Be Destroyed." *Environment and Planning D: Society and Space* 28:772–90.

Harman, Graham. 2011. *Quentin Meillassoux: Philosophy in the Making*. Edinburgh: Edinburgh University Press.

Harman, Graham. 2012. "On Interface: Nancy's Weights and Masses." In *Jean-Luc Nancy and Plural Thinking*, edited by Peter Gratton and Marie-Eve Morin, 95–107. New York: SUNY Press.

Harman, Graham. 2013. "The Current State of Speculative Realism." *Speculations* 4:22–28.

Harney, Maurita. 2015. "Naturalizing Phenomenology: A Philosophical Imperative." *Progress in Biophysics and Molecular Biology* 119 (3): 661–69.

Hawking, Stephen, and Leonard Mlodinow. 2010. *The Grand Design*. London: Bantam.

Heidegger, Martin. 1976. "Nur noch ein Gott kann uns retten." *Spiegel* 30:193–219.

Heidegger, Martin. 1981. "Only a God Can Save Us." Translated by W. Richardson. In *Heidegger: The Man and the Thinker*, edited by T. Sheehan, 45–67. Chicago: Precedent.

Heidegger, Martin. 1993. *Heidegger: Basic Writings*. Edited by David Farrell Krell. London: Routledge.

Hossenfelder, Sabine. 2007. "The Trouble with Physics Aftermath." BackRe(Action) (blog), September 19, 2007. http://backreaction.blogspot.com/.

Howells, Christina, and Gerald Moore. 2013. *Stiegler and Technics*. Edinburgh: Edinburgh University Press.

James, Ian. 2006. *The Fragmentary Demand: An Introduction to the Philosophy of Jean-Luc Nancy*. Stanford, Calif.: Stanford University Press.

James, Ian. 2010. "Bernard Stiegler and the Time of Technics." In *Cultural Politics* 6 (2): 207–28.

James, Ian. 2012. *The New French Philosophy*. Cambridge: Polity.

James, Ian. 2014. "Differing on Difference." In *Nancy Now*, edited by Verena Andermatt Conley and Irving Goh, 110–26. Cambridge: Polity.

Kant, Emmanuel. 1998. *Critique of Pure Reason*. Translated by Paul Guyer. Cambridge: Cambridge University Press.

Ladyman, James, and Don Ross. 2007. *Every Thing Must Go*. Oxford: Oxford University Press.

Ladyman, James, Don Ross, and Harold Kincaid, eds. 2013. *Scientific Metaphysics*. Oxford: Oxford University Press.

Lakatos, Imre. 1976. *Proofs and Refutations: The Logic of Mathematical Discovery*. Cambridge: Cambridge University Press.

Lane, Nick. 2015. *The Vital Question*. London: Profile Books.

Laruelle, François. 1977. *Le Déclin de l'écriture*. Paris: Aubier-Flammarion.

Laruelle, François. 1981. *Le Principe de minorité*. Paris: Aubier-Montaigne.

Laruelle, François. 1985. *Une Biographie de l'homme ordinaire. Des autorités et des minorités*. Paris: Aubier.

Laruelle, François. 1986. *Les Philosophies de la différence*. Paris: Presses Universitaires de France.

Laruelle, François. 1989. *Philosophie et non-philosophie*. Liège: Pierre Mardaga.

Laruelle, François. 1991. *En tant qu'un*. Paris: Aubier.

Laruelle, François. 1992. *Théorie des identités*. Paris: Presses Universitaires de France.

Laruelle, François. 1995. *Théorie des étrangers*. Paris, Kimé.

Laruelle, François. 1996. *Principes de la non-philosophie*. Paris: Presses Universitaires de France.

Laruelle, François. 2000. *Introduction au non-marxisme*. Paris: Presses Universitaires de France.

Laruelle, François. 2002. *Le Christ futur*. Paris: Maisonneuve Et Larose.
Laruelle, François. 2008. *Introduction aux sciences génériques*. Paris: Pétra.
Laruelle, François. 2010a. *Future Christ*. Translated by Anthony Paul Smith. London: Continuum.
Laruelle, François. 2010b. *Philosophies of Difference*. Translated by Rocco Gangle. London: Continuum.
Laruelle, François. 2010c. *Philosophie Non-standard*. Paris: Kimé.
Laruelle, François. 2011. *Anti-Badiou*. Paris: Kimé.
Laruelle, François. 2012. "La Frontière messianique." In Schmid 2012b, 303–18.
Laruelle, François. 2013a. *Anti-Badiou: On the Introduction of Maoism into Philosophy*. Translated by Robin Mackay. London: Bloomsbury.
Laruelle, François. 2013b. *Philosophy and Non-philosophy*. Translated by Taylor Adkins. Minneapolis: Univocal.
Laruelle, François. 2013c. *Principles of Non-philosophy*. Translated by Nicola Rubczak and Anthony Paul Smith. London: Bloomsbury.
Laruelle, François. 2014. *Christo-fiction*. Paris: Fayard.
Laruelle, François. 2015. *Introduction to Non-Marxism*. Translated by Anthony Paul Smith. Minneapolis: Univocal.
Laruelle, François. 2016. *Theory of Identities*. Translated by Alyosha Edlebi. New York: Columbia University Press.
Laruelle, François. 2017. *A Biography of Ordinary Man: On Authorities and Minorities*. Translated by Jessie Hock and Alex Dubilet. Cambridge: Polity.
Leblanc, Guillaume. 2002. *La Vie Humaine. Anthropologie et biologie chez Georges Canguilhem*. Paris: Presses Universitaires de France.
Lewis, David. 1986. *On the Plurality of Worlds*. Oxford: Blackwell.
Lewis, David. 1999. *Papers in Metaphysics and Epistemology*. Cambridge: Cambridge University Press.
Longo, Giuseppe, and Francis Bailly. 2009. "Biological Organization and Anti-entropy." *Journal of Biological Systems* 17 (1): 63–96.
Longo, Giuseppe, and Maël Montévil. 2014. *Perspectives on Organisms: Biological Time, Symmetries and Singularities*. Berlin: Springer Verlag.
Malabou, Catherine. 1996. *L'Avenir de Hegel: Plasticité, temporalité, dialectique*. Paris: J. Vrin.
Malabou, Catherine. 2004a. *Le Change Heidegger: Du Fantastique en philosophie*. Paris: Éditions Léo Scheer.
Malabou, Catherine. 2004b. *Que faire de notre cerveau?* Paris: Bayard.
Malabou, Catherine. 2005a. *The Future of Hegel*. Translated by Lisabeth During. London: Routledge.
Malabou, Catherine. 2005b. *La Plasticité au soir de l'écriture*. Paris: Éditions Léo Scheer.
Malabou, Catherine. 2008. *What Should We Do with Our Brain?* Translated by Sebastien Rand. New York: Fordham University Press.
Malabou, Catherine. 2010. *Plasticity at the Dusk of Writing*. Translated by Carolyn Shread. New York: Columbia University Press.

Malabou, Catherine. 2011. *The Heidegger Change*. Translated by Peter Skafish. New York: SUNY Press.

Malabou, Catherine. 2014. *Avant demain: Épigenèse et rationalité*. Paris: Presses Universitaires de France.

Malabou, Catherine. 2016. *Before Tomorrow: Epigenesis and Rationality*. Translated by Carolyn Shread. Cambridge: Polity.

Malabou, Catherine. 2017. *Métamorphoses de l'intelligence: Que faire de leur cerveau bleu?* Paris: Presses Universitaires de France.

Mathiot, Jean. 1993. "Génétique et connaissance de la vie." In *Georges Canguilhem: Philosophe, historien des sciences*, 194–207. Paris: Albin Michel.

Maturana, Humberto, and Francisco Varela. 1980. *Autopoiesis and Cognition: The Recognition of the Living*. Dordrecht: Reidel.

Mazzocchi, Fulvio. 2008. "Complexity in Biology." *EMBO Reports* 9 (1): 10–14.

Meillassoux, Quentin. 2010. *After Finitude: An Essay on the Necessity of Contingency*. London: Continuum.

Mitchell, Sandra D. 2003. *Biological Complexity and Integrative Pluralism*. Cambridge: Cambridge University Press.

Mullarkey, John. 2006. *Post-Continental Philosophy: An Outline*. London: Continuum.

Mullarkey, John. 2012. "I + I = I: The Non-consistency of Non-philosophical Practice (Photo: Quantum: Fractal)." In *Laruelle and Non-philosophy*, edited by John Mullarkey and Anthony Paul Smith, 143–68. Edinburgh: Edinburgh University Press.

Mulqueen, Tara, and Daniel Matthews, eds. 2015. *Being Social: Ontology, Law, Politics*. Oxford: Counterpress.

Nancy, Jean-Luc. 1976. *Logodeadalus: Le Discours de la syncope*. Paris: Aubier-Flammarion.

Nancy, Jean-Luc. 1979. *Ego sum*. Paris: Flammarion.

Nancy, Jean-Luc. 1983. *L'Impératif catégorique*. Paris: Flammarion.

Nancy, Jean-Luc. 1986. *La Communauté désœuvrée*. Paris: Christian Bourgois.

Nancy, Jean-Luc. 1988. *L'Expérience de la liberté*. Paris: Galilée.

Nancy, Jean-Luc. 1990. *Une Pensee finie*. Paris: Galilée.

Nancy, Jean-Luc. 1991. *The Inoperative Community*. Translated by Peter Connor et al. Minneapolis: University of Minnesota Press.

Nancy, Jean-Luc. 1992. *Corpus*. Paris: Métaillé.

Nancy, Jean-Luc. 1993a. *The Birth to Presence*. Translated by B. Holmes et al. Stanford, Calif.: Stanford University Press.

Nancy, Jean-Luc. 1993b. *The Experience of Freedom*. Translated by B. McDonald. Stanford, Calif.: Stanford University Press.

Nancy, Jean-Luc. 1993c. *Le Sens du monde*. Paris: Galilée.

Nancy, Jean-Luc. 1996. *Être singulier pluriel*. Paris: Galilée.

Nancy, Jean-Luc. 1997. *The Sense of the World*. Translated by Jeffrey S. Librett. Minneapolis: University of Minnesota Press.

Nancy, Jean-Luc. 2000. *Being Singular Plural*. Translated by Robert Richardson and Anne O'Byrne. Stanford, Calif.: Stanford University Press.

Nancy, Jean-Luc. 2007. *The Discourse of the Syncope: Logodeadalus*. Translated by Saul Anton. Stanford: University Press.

Nancy, Jean-Luc. 2016. *Ego Sum*. Translated by Marie-Eve Morin. New York: Fordham University Press.

Nancy, Jean-Luc, and Aurélien Barrau. 2011. *Dans quels mondes vivons-nous?* Paris: Galilée.

Nancy, Jean-Luc, and Aurélien Barrau. 2014. *What's These Worlds Coming To?* Translated by Travis Holloway and Flor Méchain. New York: Fordham University Press

Ó Maoilearca, John. 2015. *All Thoughts Are Equal: Laruelle and Nonhuman Philosophy*. Minneapolis: University of Minnesota Press.

Olafson, Frederick A. 2001. *Naturalism and the Human Condition: Against Scientism*. London: Routledge.

Papineau, David. 1993. *Philosophical Naturalism*. Oxford: Blackwell.

Pashler, Harold, Christine Harris, Piotr Winckelman, and Edward Vul. 2009. "Puzzlingly High Correlations in fMRI Studies of Emotion, Personality, and Social Cognition." *Perspectives on Psychological Science* 14 (3): 274–90.

Penrose, Roger. 2016. *Fashion, Faith and Fantasy in the New Physics of the Universe*. Princeton, N.J.: Princeton University Press.

Poincaré, Henri. (1902) 2016. *Science and Hypothesis*. Translated by W. J. Greenstreet. London: Read Books.

Polchinski, Joseph. 2007. "All Strung Out?" *American Scientist* 95 (1): 72–75.

Prigogine, Ilya, and Isabelle Stengers. 1988. *Entre le temps et l'éternité*. Paris: Fayard.

Protevi, John. 2013. *Life, War, Earth: Deleuze and the Sciences*. Minneapolis: University of Minnesota Press.

Quine, W. V. 1953. *From a Logical Point of View*. Cambridge, Mass.: Harvard University Press.

Quine, W. V. 1966. *The Ways of Paradox*. Cambridge, Mass.: Harvard University Press.

Quine, W. V. 1981. *Theories and Things*. Cambridge, Mass.: Harvard University Press.

Quine, W. V. 1990a. "Naturalism, or Living within One's Means." *Dialectica* 49:251–61.

Quine, W. V. 1990b. *The Pursuit of Truth*. Cambridge, Mass.: Harvard University Press.

Reydon, Thomas. 2010. "How Special Are the Life Sciences: A View from the Natural Kinds Debate." In *The Present Situation in the Philosophy of Science*, edited by Friedrich Stadler, 173–88. Berlin: Springer.

Rosenberg, Alexander. 2014. "Disenchanted Naturalism." In Bashour and Muller 2014, 13–17.

Rouvroy, Antoinette, and Thomas Berns. 2013. "Gouvernmentalité algorithmique et persectives d'émancipation." Réseaux 177:163–96.

Rovelli, Carlo. 1996. "Relational Quantum Mechanics." *International Journal of Theoretical Physics* 35 (8): 1637–78.

Rovelli, Carlo. 2014. *Reality Is Not What It Seems: The Journey to Quantum Gravity*. Harmondsworth: Penguin.

Sankey, Howard. 2004. "Scientific Realism: An Elaboration and a Defence." In Carrier, Roggenhofer, Küppers, and Blanchard 2004, 55–79.

Schmid, Anne-Françoise. 2001. *Que peut la philosophie des sciences?* Paris: Éditions Pétra.

Schmid, Anne-Françoise. 2012a. "Introduction: Sciences et philosophies, la question des frontières." In Schmid 2012b, 13–39.

Schmid, Anne-Françoise, ed. 2012b. *Épistémologie des frontières*. Paris: Éditions Pétra.

Schmid, Anne-Francoise, and Nicole Mathieu, eds. 2014. *Modélisation et interdisciplinarité: Si disciplines en quête d'épistémologie*. Paris: Éditions Quæ.

Schrödinger, Erwin. (1944) 1962. *What Is Life?* Cambridge: Cambridge University Press.

Sellars, Wilfrid. 1997. *Empiricism and the Philosophy of Mind*. Cambridge, Mass.: Harvard University Press.

Shannon, C. 1948. "A Mathematical Theory of Communication." *Bell System Technical Journal* 27:379–423, 623–56.

Shannon, C. E., and W. Weaver. 1949. *The Mathematical Theory of Communication*. Urbana: University of Illinois Press.

Shaviro, Steven. 2014. *The Universe of Things: On Speculative Realism*. Minneapolis: University of Minnesota Press.

Sklar, Lawrence. 2003. "Dappled Theories in a Uniform World." *Philosophy of Science* 70:424–41.

Smith, Anthony Paul. 2013. *A Non-philosophical Theory of Nature: Ecologies of Thought*. Basingstoke: Palgrave Macmillan.

Smith, Anthony Paul. 2016a. *François Laruelle's* Principles of Non-philosophy*: A Critical Introduction and Guide*. Edinburgh: Edinburgh University Press.

Smith, Anthony Paul. 2016b. *Laruelle*. Cambridge: Polity.

Smolin, Lee. 1997. *The Life of the Cosmos*. Oxford: Oxford University Press.

Smolin, Lee. 2007. *The Trouble with Physics*. London: Allen Lane.

Smolin, Lee. 2013. *Time Reborn*. London: Allen Lane.

Sparrow, Tom. 2014. *The End of Phenomenology: Metaphysics and the New Realism*. Edinburgh: Edinburgh University Press.

Stengers, Isabelle. 1993. *L'Invention des sciences modernes*. Paris: Éditions de la Découverte.

Stengers, Isabelle. 2000. *The Invention of Modern Science*. Translated by Daniel W. Smith. Minneapolis: University of Minnesota Press.

Stengers, Isabelle. 2002. *Penser avec Whitehead: Une libre et sauvage création de concepts*. Paris: Seuil.

Stengers, Isabelle. 2003. *Cosmopolitiques 1*. Paris: Éditions de la Découverte.

Stengers, Isabelle. 2010. *Cosmopolitics 1*. Translated by Robert Bononno. Minneapolis: University of Minnesota Press.

Stengers, Isabelle. 2011. *Thinking with Whitehead: A Free and Wild Creation of Concepts*. Translated by Michael Chase. Cambridge, Mass.: Harvard University Press.

Stiegler, Bernard. 1994. *La Technique et le temps 1. La faute de l'Épiméthée*. Paris: Galilée.

Stiegler, Bernard. 1998. *Technics and Time 1. The Fault of Epimetheus*. Translated by Richard Beardsworth and George Collins. Stanford, Calif.: Stanford University Press.

Stiegler, Bernard. 2001. "Derrida and Technology." In *Jacques Derrida and the Humanities*, edited by Tom Cohen, 238–70. Cambridge: Cambridge University Press.

Stiegler, Bernard. 2003. *Passer à l'acte*. Paris: Galilée.

Stiegler, Bernard. 2009. *Acting Out*. Translated by David Barison, Daniel Ross, and Patrick Crogan. Stanford, Calif.: Stanford University Press.

Stiegler, Bernard. 2012. *États de choc. Bêtise et savoir au XXIe siècle*. Paris: Millet et Une Nuits.

Stiegler, Bernard. 2015a. *La Société automatique 1. L'Avenir du travail*. Paris: Fayard.

Stiegler, Bernard. 2015b. *States of Shock: Stupidity and Knowledge in the 21st Century*. Cambridge: Polity.

Stiegler, Bernard. 2016a. *The Automatic Society 1: The Future of Work*. Translated by Daniel Ross. Cambridge: Polity.

Stiegler, Bernard. 2016b. *Dans la disruption. Comment ne pas devenir fou?* Paris: Éditions Les Liens qui Libèrent.

Timpson, Christopher G. 2013. *Quantum Information Theory and the Foundations of Quantum Mechanics*. Oxford: Oxford University Press.

Unger, Roberto Mangabeira, and Lee Smolin. 2015. *The Singular Universe and the Reality of Time: A Proposal in Natural Philosophy*. Cambridge: Cambridge University Press.

Varela, Francisco. 1991. *The Embodied Mind: Cognitive Science and Human Experience*. Cambridge, Mass.: MIT Press.

Wallace, David. 2012. *The Emergent Multverse*. Oxford: Oxford University Press.

Wheeler, John Archibald. 1999. "Information, Physics, Quantum: The Search for Links." In *Feynman and Computation: Exploring the Limits of Computers*, edited by Anthony J. G. Hey, 309–36. Reading, Mass.: Perseus.

Whitehead, Alfred North. (1920) 2015. *The Concept of Nature*. Cambridge: Cambridge University Press.

Wigner, Eugene. 1960. "The Unreasonable Effectiveness of Mathematics in the Natural Sciences." *Communications in Pure and Applied Mathematics* 13 (1): 1–14.

Wolfendale, Peter. 2014. *Object-Oriented Philosophy: The Noumenon's New Clothes*. Falmouth: Urbanomic.

Worral, John. 1989. "Structural Realism: The Best of Both Worlds?" *Dialectica* 43:99–124.

Wrathall, Mark, and Sean Kelly. 1996. "Existential Phenomenology and Cognitive Science." *Electronic Journal of Analytic Philosophy*, no 4. https://ejap.louisiana.edu/EJAP/1996.spring/wrathall.kelly.1996.spring.html.

INDEX

IAN JAMES is a fellow of Downing College and a reader in modern French literature and thought in the faculty of modern and medieval languages at the University of Cambridge. He is the author of *Pierre Klossowski: The Persistence of a Name, The Fragmentary Demand: An Introduction to the Philosophy of Jean-Luc Nancy, Paul Virilio,* and *The New French Philosophy.*